Spaces for the Sacred

Spaces for the Sacred

Place, Memory, and Identity

Philip Sheldrake

The Johns Hopkins University Press
Baltimore, Maryland

This edition first published in 2001 by SCM Press, London.
Johns Hopkins Paperbacks edition, 2001
2 4 6 8 9 7 5 3 1

The Johns Hopkins University Press
2715 North Charles Street
Baltimore, Maryland 21218-4363
www.press.jhu.edu

Library of Congress Cataloging-in-Publication Data

Sheldrake, Philip.
Spaces for the sacred : place, memory, and identity / Philip Sheldrake.
p.   cm.
Includes bibliographical references (p.) and index.
ISBN 0-8018-6861-0 (pbk. : alk. paper)
1. Sacred space. 2. Spiritual life—Christianity. I. Title.
BV895 .S54 2001
263′.042—dc21   2001038266

A catalog record for this book is available from the British Library.

Contents

Preface

The gestation period of this book was a long one. I was privileged to be appointed the Hulsean Lecturer in the University of Cambridge for 1999–2000 and so the immediate origins of the book lie in the lectures I gave in the Divinity Faculty on 'Place, Memory and Human Identity'. The book is substantially based on those lectures. I am very grateful to the Electors for giving me the opportunity to gather together a number of scattered thoughts on the topic of place – a subject that has fascinated me for many years. I also wish to thank the Regius Professor, David Ford, and other members of the Faculty for their warm hospitality to me and to my wife Susie during the weekly visits and to thank Rosalind Paul for the many practical arrangements she made on our behalf.

It would be strange indeed if a book on place were to be entirely unplaced and so a modest amount of personal detail seems entirely appropriate. I first became interested in a sense of place during my childhood, long before I would have spoken in those terms. I grew up in a landscape and on a coastline that echoed with millennia of human associations, including a succession of invasions (Roman, Saxon and Norman). The result was that I became attuned at an early age not merely to the beauties of countryside and seascape but also to the relationship between place, memory and our human narratives. Then as an adult I spent a year studying in India. It was my first experience of living outside Western culture. That led me to reflect more deeply not only on the importance of roots, of a sense of belonging 'somewhere', but also on the transformative power of extended journeys away from home to an elsewhere.

The first time I consciously began to address the theme of place was in a series of conversations during a visit to Australia in 1990.

Later still, during a university sabbatical in 1992, I spent time studying and writing for several months in three strikingly different and powerful places. The first was the medieval monastic landscape of North Yorkshire where I began to reread the classic monastic rules. The second was the small island of Columba's Iona jutting out into the wildness of the Atlantic where I immersed myself in some of the texts of Celtic Christianity. Finally there was a small Benedictine monastery buried deep in the canyon solitudes of the New Mexican desert. There, among other things, I began to read contemporary studies on place and to dip into modern American literature which is so preoccupied with the theme. Although this was not my original intention, the immediate result of the sabbatical was a short book on Celtic Christianity with a particular focus on place and journey. That sabbatical also began what turned into a programme of extensive reading on the theme of place that was interdisciplinary but with a theological end in view.

Here I must thank colleagues and students for guiding my reading and stimulating my reflections. Laurence Cunningham, then Chair of the theology department at the University of Notre Dame, invited me in 1995 to give a public lecture dedicated to the theme of place. That ended up as a chapter in an earlier book and provoked a great deal of interest. Successions of students in the Cambridge Theological Federation in England and at Notre Dame had to suffer from, and contributed greatly to, my search for a deeper understanding of the subject. Over these years I have benefited from many conversations about place from different perspectives with friends and colleagues in the UK and the USA. Place appears to be a topic guaranteed to provoke energy and interest. However, I must record a particular debt of thanks to two friends and American colleagues in the Society for the Study of Christian Spirituality. Doug Burton-Christie and Belden Lane have not only actively encouraged my own research by sending me articles or by pointing me to references but have also stimulated me by their own important writing on the topic (different in perspective from my own) and by lively conversation.

Throughout this book I attempt to suggest a number of specifically Christian perspectives from which we may think more creatively about place. My overall approach is from within the interdisciplinary field of spirituality but with a particular concern for its relationship with theology. However, my claim is rather modest. The book is intended to be an exploration of ideas and interpretative perspectives rather than a systematic study of place or an attempt to produce a comprehensive thesis on the subject. I merely suggest a number of starting-points that I hope will stimulate further creative questions. For one thing, in terms of a theological project I have not attempted to survey the considerable amount of complex biblical material. I am also aware that I have given only slight attention to the way place has become an increasingly important theme in contemporary literature in the English-speaking world. Finally, I have postponed as premature any attempt to enter into the debates concerning place in discussions about the Internet and virtual reality. Despite these obvious gaps, I hope that what follows will none the less be useful and provocative.

The first three chapters offer some foundations for the discussion. In Chapter 1, I suggest some building blocks for a fundamental definition of place as a human construct. Chapter 2 casts a more historical gaze on the way that Christianity turned its attention away from geography towards people as *loci* of the sacred. Chapter 3 suggests some theological trajectories in terms of the coinherence of sacramental and ethical perspectives. The other three chapters concern what one might call 'the practice of place'. Chapter 4 examines the importance of imaginary or utopian (even possibly heterotopian) place in relation to the monastic tradition. Chapter 5 explores the impulse within Christian mysticism to move beyond 'place-as-locality' in themes of journey and perpetual departure. The final chapter concentrates on the contemporary issue of an 'integral ecology' of the human city and invites us to explore a more positive vision of its future meaning.

I

A Sense of Place

Our preoccupation with the turn of the millennium has once again focused attention on the categories of human experience that have the greatest impact on the way we see the world. Apart from time, the controversy over the building of the Millennium Dome in wasteland near Greenwich has also highlighted the strength of our feelings about place. If the structure of the Dome and its contents is a kind of text that reflects the ways in which contemporary British culture operates, what does this imply? As the only national 'building' constructed in the United Kingdom to mark the millennium, the Dome has been compared unfavourably as an expression of human creativity with what our ancestors built in the past. Simply as a construction it has no long-term existence. In terms of content, it is orientated to the present moment only and offers little sense of the history that has brought us to this time and that points forward into our future.

The misgivings about the Dome, justified or not, highlight the vital connection between three things: place, memory and human identity. The concept of place refers not simply to geographical location but to a dialectical relationship between environment and human narrative. Place is space that has the capacity to be remembered and to evoke what is most precious. 'We need to think about where we are and what is unique and special about our surroundings so that we can better understand ourselves and how we relate to others.'[1] Because of this, the human sense of place is a critical theological and spiritual issue. Yet, historically, the Christian tradition has been ambivalent about the subject. This partly reflects a tension between place and placelessness that goes back to the biblical origins of Christian faith. The unfortunate result, however,

is that in current debates about the future of place, the Christian
theological voice contributes very little apart from occasional references to specifically environmental issues.

A crisis of place?

Place has become a significant theme in a wide range of writing
including philosophy, cultural history, anthropology, human geography, architectural theory and contemporary literature. This
partly reflects what a number of commentators refer to as a crisis of
place in Western societies – a sense of rootlessness, dislocation or
displacement. Part of this crisis is cultural. At its root lies a decline
in traditional systems of values and symbols – religious, ethical and
social. The resulting fragmentation tends, among other things, to
inhibit a clear world-view. This new context, often labelled 'post-
modernity', has turned its back on the intellectual and cultural
optimism of the 'modern' period that had held sway since the
European Enlightenment and the Industrial Revolution. Despite
today's rapid technological developments, it is no longer possible to
believe in the absolute inevitability of progress or in the capacity of
reason alone to grasp the meaning of everything. People are
increasingly suspicious of universal frameworks of meaning that
claim to possess 'the truth' in some total way.

In this context, the present intellectual experience of place has
been characterized as a movement through a wasteland among the
ruins of former theories of meaning. In the uncertainties of a postmodern world a culture of individual choice, especially the freedom
to choose from the widest possible range, tends to take over from
social reconstruction as the goal towards which societies strive.
Even the increasing interest in spirituality is often individualistic
and privatized. Some people suggest that the post-Marxist death of
teleological theories of history leaves us to exist in an 'unspecific
intellectual location' that leads precisely 'nowhere'.[2] The question
is whether this is actually a place of transition towards an eventual,
even if more modest, recovery of some kind of collective meaning.
One of the outstanding French thinkers of the late twentieth century, and sometime Jesuit, the late Michel de Certeau, offered what

might be called a more spiritual interpretation of postmodern place. Influenced in strange ways to the end of his life by what might be called 'the Jesuit project', de Certeau suggested that we are on a kind of perpetual pilgrimage that somehow parallels the mystical tradition. We experience dissatisfaction with final definitions or completed places and are driven ever onwards in a movement of perpetual departure. Each of us 'with the certainty of what is lacking, knows of every place and object that it is *not that*, one cannot stay *there* nor be content with *that*'.[3]

Culture and world-views

We not only live in the world but also have an image or picture of the world. People in any society exist within a system of signs through which they identify themselves and understand their world. This process is conditioned in part by fundamental beliefs about God or about the nature of the human condition. The world that surrounds them is not simply raw data but something they experience as bearing meaning. Indeed, the very notion of the world is a human construct. We do not dwell in pure nature but in 'the realm of mediated meaning'.[4] This system of meaning or world-view is what anthropologists nowadays understand by the term 'culture'. Culture 'denotes an historically transmitted pattern of meanings embodied in symbols, a system of inherited conceptions expressed in symbolic forms by means of which men [*sic*] communicate, perpetuate, and develop their knowledge about and attitudes toward life'.[5] This semiotic approach emphasizes that culture is a text, potentially with many layers of meaning. It demands sophisticated reading and complex interpretation rather than relatively straightforward classification and explanation. '[A]s interworked systems of construable signs . . . culture is not a power, something to which social events, behaviours, institutions or processes can be causally attributed; it is a context, something within which they can be intelligibly – that is, thickly – described'.[6] The symbols, rituals, attitudes and perspectives about life that constitute 'culture' enable human societies to cohere and function. Culture regulates how people assign meaning and allocate value in terms of the key elements of human living. It defines their social, economic, political and religious behaviour:

> Human society is in permanent motion, change and development. At different times and in different cultures men perceive and interpret the world in their own fashion, and in their own fashion they organise their impressions and their knowledge, and construct their own historically conditioned worldview.[7]

Without becoming involved in technical discussions about cultural theory, we need to consider some recent shifts of understanding that will be relevant throughout this book. Essentially the older assumption that the world is simply a mosaic of separate cultures is now questionable. On the one hand, technology, rapid travel and contemporary economic processes have produced a rapid increase in regional and global connections. Place is no longer simply local. On the other hand, when we do turn our attention to the cultures associated with particular peoples in local places, the focus is no longer exclusively on culture as essentially an ordered phenomenon. Cultures previously viewed as homogeneous are now revealed as plural, often fragmented and inextricably associated with issues of partiality, dominance and exclusions – in other words, with power. All this has a significant impact on our perceptions of place, and any theological reflections on it – precisely because place has emerged as a contested reality.[8]

Place is a cultural category

Place, like time, is among the most universal of cultural categories although it clearly operates in different ways in each specific context. Few factors in a culture express its world picture so clearly, because place has a determining influence on the way people behave, the way they think, the rhythm of their lives and their relationships.[9] Physical places are vital sources of metaphors for our social constructions of reality. Metaphors are not optional extras or embellishment to our normal ways of thinking and speaking. Our perceptions of reality are defined by metaphor:[10] 'Places form landscapes and landscapes may be defined as sets of relational places each embodying (literally and metaphorically) emotions, memories and associations derived from personal and interpersonal shared experience.'[11]

The most prolific contributions to contemporary understandings of place stem from the work of cultural geographers and anthropologists. These in turn mainly rely on philosophical concepts derived from the phenomenology of Martin Heidegger (such as the concept of dwelling) or from postmodern thinkers such as Michel de Certeau (such as the question of belonging). In recent years writers on place have turned their attention to social identity in relation to roots and rootlessness and, provoked by various forms of displacement, to the politics of place, especially to what are referred to as 'geographies of struggle and resistance'.

In terms of religious anthropology, figures such as Mircea Eliade and Victor and Edith Turner (strongly influenced by Eliade's theories) dominated the field of studies about place and the sacred until recently. They tended to assume that both sacred places and pilgrimage could be understood in a single way. For Eliade, every sacred place was to be thought of as an *axis mundi*, the centre of the world, with boundaries separating it from surrounding secular or profane space. Such places were a kind of Jacob's ladder linking heaven and earth. Building on this universal theory, the Turners suggested in their discussions of pilgrimages and sacred places that these created what they called *communitas*. This was a special temporary state in which conventional social or other distinctions are transcended in a spontaneous sharing of experience.

However, more recent writing on the subject is much more sensitive to the *plurality* of meanings given to particular places by those who relate to them. Places in general, whether sacred or not, do not have a single 'given' meaning but are interpreted quite differently by distinct groups of people. Indeed, as the very title of a recent volume of essays on pilgrimage (*Contesting the Sacred*) suggests, place and the sacred are just as likely to cause division as provoke consensus and harmony. A classic example would be Jerusalem and the Holy Land in which no fewer than three major world faiths (and sub-groups within them) contest the meaning of sacred sites.[12] In general, discussions about place until recently were not much concerned with such political questions. There was a lack of attention to the contentious issues of the loss of roots and the experience of being uprooted in exile or as refugees. However, the subject of contested place and the link between place and power is nowadays a major area of study.[13] I will return to this subject later.

Place and space

The meanings of places unfold in stories, myths, rituals and in naming. The social significance of places finds expression in music, art and architecture. Philosophers and others who reflect on place have moved away from the notion that empty space is the fundamental, natural reality and that place is a secondary, albeit necessary, social construction that gives meaning to what is otherwise a *tabula rasa*.

Older scientific views of reality suggested the priority of space as absolute, infinite, empty and *a priori*. Place (or more accurately, a plurality of places) was a mere apportioning or compartmentalization of 'natural' space. This view has proved problematic in several ways. First of all, it suggests that there really is such an objective reality as nature apart from how we interpret it. This is associated with an intellectual preference for the universal or the general over the local or the particular and for abstract definition and objective knowledge rather than for what derives from experience. Second, such a view makes nature a morally neutral reality on which we can impose whatever we choose. Such intellectual preferences also tend to marginalize particular stories in favour of a single overarching narrative. The issue is not merely philosophical but also political, ethical and, in the end, spiritual.

Third, the notion of space as three-dimensional, geometrical, evenly extended and divisible into commensurate sections has been complicated by the theory of relativity, developments in particle physics and even in the psychology of perception. Space does not exist as an objective 'thing'. It is subjectively perceived and experienced differently depending on perspective. For example, space can now be said to be compressed. New methods of communication and travel enable us to span great distances and a greater distance per time unit than we could achieve even twenty-five years ago. In that sense, the world may be described as smaller. Theorists are now much more inclined to say that, if it is appropriate to speak of priorities, a sense of *place* actually precedes and creates a sense of space. Space is an abstract analytical concept whereas place is always tangible, physical, specific and relational.

Philosophers such as Martin Heidegger, Gaston Bachelard and Edward S. Casey have re-embraced the conviction that place is prior to space. We come to know in terms of the particular knowledge of specific places before we know space as a whole or in the abstract.[14] 'Spaces receive their being from locations and not from "space".'[15] In his essay 'An Ontological Consideration of Place', Heidegger insisted that 'place is the house of being'.[16] 'To say that mortals *are* is to say that *in dwelling* they persist through spaces by virtue of their stay among things and locations.'[17] A 'person' for Heidegger was *Dasein*, or 'being-there'. In other words, to be a person is literally 'to be there', to be in a particular place.

Walter Brueggemann in his bold and controversial work on 'the Land' as a central theme of biblical faith underscores the important distinction between space and place and argues that it is within the spatial connections of human life that we most deeply encounter the meaning of existence:

> Place is space which has historical meanings, where some things have happened which are now remembered and which provide continuity and identity across generations. Place is space in which important words have been spoken which have established identity, defined vocation and envisioned destiny. Place is space in which vows have been exchanged, promises have been made, and demands have been issued. Place is indeed a protest against an unpromising pursuit of space. It is a declaration that our humanness cannot be found in escape, detachment, absence of commitment, and undefined freedom . . . Whereas pursuit of space may be a flight from history, a yearning for a place is a decision to enter history with an identifiable people in an identifiable pilgrimage.[18]

Place depends on relationships and memories as much as on physical features. It is 'a complex network of relationships, connections and continuities . . . of physical, social and cultural conditions that describe my actions, my responses, my awareness and that give shape and content to the very life that is me'.[19]

Place and social crisis

Another dimension of the contemporary crisis of place is social. People in the West are increasingly an exiled and uprooted people, living 'out of place'. Social geographers suggest that while it is essential to have 'place identity', we have since the Second World War de-emphasized place for the sake of values such as mobility, centralization or economic rationalization. The global relativity of space dissolves a human sense of place. 'The skyscrapers, airports, freeways and other stereotypical components of modern landscapes – are they not the sacred symbols of a civilisation that has deified reach and derided home?'[20] Indeed, mobility is now understood to be a freedom bought by money and education. Remaining in the same place has come to symbolize a lack of choice, an entrapment, which is the lot of the poor, the elderly and people with handicaps. In an increasingly placeless culture we become 'standardised, removable, replaceable, easily transported and transferred from one location to another'.[21]

The French anthropologist Marc Augé has described what he terms 'non-place'. He distinguishes between place, full of historical monuments and creative of social life, and non-place where no organic social life is possible. By non-place Augé means the contexts where we spend more and more time such as supermarkets, airports, hotels, motorways, in front of the television, sitting at a computer and so on. These experiences bring about a fragmentation of awareness that leads to incoherence in relation to 'the world'. Augé describes non-place as 'curious places which are both everywhere and nowhere'. By contrast, place is a concrete and symbolic construction of space that serves as a reference for all those to whom it assigns a position. Place is also a principle of meaning for those who live in it and a principle of intelligibility for those who observe it. Unlike non-place, place has three essential characteristics – it engages with our identity, with our relationships and with our history.[22]

Media and technology are also involved in contemporary place-lessness, or deterritorialization as some social theorists call it.[23] Rather than the 'global village' with its strong communitarian imagery of locality, media and information technology are just as likely to create communities with no sense of place. In a 'dramatically delocalized world' what is locality? It 'seems to have lost its ontological moorings'.[24]

Place and belonging

It is this sense of placelessness that makes the contemporary Western quest for meaning so concerned with roots. 'There are no meanings apart from roots.'[25] Our longing for place is more than biological or aesthetic. Simone Weil suggested that the hunger for roots is fundamental to our deepest identity:

> To be rooted is perhaps the most important and least recognised need of the human soul. It is one of the hardest to define. A human being has roots by virtue of his real, active and natural participation in the life of a community which preserves in living shape certain particular expectations for the future. This participation is a natural one, the sense that it is automatically brought about by place, conditions of birth, profession and social surroundings. Every human being needs to have multiple roots. It is necessary for him to draw well-nigh the whole of his moral intellectual and spiritual life by way of the environment of which he forms a natural part.[26]

The most fundamental fact of human existence is that because people are embodied they are always 'somewhere'. In the West, most people are housed and so take for granted that they have a defined location and an address in some human environment. This expresses, more or less, Heidegger's idea of dwelling. Place is the possibility of genuine dwelling. 'We attain to dwelling, so it seems, only by means of building'.[27] Yet, as Heidegger himself recognized, buildings in themselves do not necessarily guarantee that dwelling happens in them. To dwell certainly involves being housed in some sense. However, to dwell is more than a matter of good planning or the design of human environments, important though these are.

We also seek an authentic place in a number of other ways. A sense of 'home' seems to be vital if human identities are not to be dispersed and fragmented. Is it simply the native, original place – literally the place of birth? *The Poetics of Space* by the French philosopher Gaston Bachelard is one of the most influential books on 'home'. He suggests, 'For our house is our corner of the world. As has often been said, it is our first universe, a real cosmos in every sense of the word.' But 'home' is more than simply where we originate. 'All really inhabited space bears the essence of the notion of home.'[28] Bachelard's emphasis on home undoubtedly reflects a tendency in the West since the nineteenth century to idealize domesticity as the shaping symbol of a satisfactory life. I will return to this in Chapter 4 when I examine the monastic asceticism of counter-domesticity. However, even if we agree that specific cultural and historical factors shape the strong emphasis on 'home', 'dwelling' and 'roots' in the writings of, for example, Heidegger, Bachelard and Weil, these concepts nevertheless represent something critical about spatial experience. First, 'home' stands for the fact that we persistently need a location where we can pass through the stages of life and become the person we are potentially. Second, we need a place where we can belong to a community. Third, we need a place that offers a fruitful relationship with the natural elements, with plants and animals and with the rhythms of the seasons. Finally, we need a place that offers access to the sacred (however we understand that term) – perhaps, crucially, relates us to *life itself as sacred*.[29]

'Belonging' involves both a connection to specific places and also our existence within networks of stable relationships. In European terms, until recently the parish was the boundary of many people's world. This was both a geographical and social reality. There were inextricable links between where you came from and who you were. The parish tended to dominate other human associations. People belonged to it from birth to death and beyond. The ancestors were already in the churchyard and present inhabitants would doubtless be buried there in their turn. This sense of place, shaped by social and religious ties as well as landscape, was intense. Even the next valley was 'other', strange and foreign. People felt spiritually and humanly dislocated when they moved, voluntarily or not, beyond their familiar boundaries. The parish even determined not merely the behaviour of people who belonged to it but equally how they thought and felt – or did *not* feel, as the case may be. It is recorded that one man remained completely untouched when a whole church full of people wept over a particular sermon. When asked why he alone had not cried, he looked surprised and replied, 'But I'm from another parish.'[30]

The fact that identity has traditionally been so strongly 'placed' partly explains why travel is such an ambiguous reality. 'Well-travelled' may be a metaphor for wisdom and moral authority. On the other hand, the people called 'travellers', the gypsies, *gens de voyage*, are amongst the most feared and disliked of people in much of Western Europe. The suspicion of professional travellers also explains in part why 'being in trade' as opposed to holding land was viewed as socially inferior. The trader is one who crosses boundaries and is therefore an alien and stranger wherever he or she goes. Recent studies of the history of European anti-semitism, particularly as it developed during the Middle Ages, reinforce this understanding. They reveal that one element in the widespread antipathy to Jews was that they were especially prominent as merchants and traders. This relates to a number of factors, including supposedly theological-ethical ones associated with commerce. However, what seems to have been equally significant, if not more so, was that Jews came to symbolize the classic 'strangers' who intruded into stable, fixed locations and disturbed their inhabitants. They had no fixed place in the social environment. They might, individually, be people of wit and charm but they had no organic connections with established social frameworks through ties of kinship, place or role.[31]

Place and commitment

Place thus has a great deal to do with commitment to human contexts and being accepted within them. Some recent writing on the psychology of place speaks of *participation* as a key element in being effectively placed. A place, as opposed to a location, a mere object 'over there', invites participation in an environment. 'Environment', in the full sense, implies different sets of relationships both between people and between the natural habitat and human beings. The psychologist David Canter suggests a three-fold model of place:

> A place is the result of relationships between actions, conceptions and physical attributes. It follows that we have not fully identified the place until we know a. what behaviour is associated with, or it is anticipated will be housed in, a given locus, b. what the physical parameters of that setting are, and c. the descriptions, or conceptions, which people hold of that behaviour in that physical environment.[32]

Commitment is the necessary corollary of participation. In Penelope Lively's recent novel, *Spiderweb*, Stella Brentwood is a cultural anthropologist who retires to settle in a Somerset village after years of wandering the world with no fixed abode. Coming to terms with 'home' and making sense of place are central themes of the book:

> 'I hope the new home is up to expectations,' Richard had said just now, and for an instant she hadn't understood what on earth he was on about. Whose home? Ah – her home, of course. This was what she now had, apparently. And must set to and play the part. Nest. Embellish. Fix rogue radiators, fit washers to taps.[33]

Really *being* somewhere means to be committed to a place rather than simply an observer. But this not what Stella is used to. For her, 'The world is out there, richly stocked and inviting observation' (p. 15). She has no difficulty in appreciating the theory that place is complex and is far more than landscape:

And thus Stella learned. There came beams of light. The place took shape. It ceased to be a landscape, a backdrop, and became an organism. Stella perceived the intricate system of checks and balances by which things worked. She saw that there was a continuous state of negotiation, of dealing, of to-and-fro arrangements. Everyone stood in a particular relationship to everyone else, often literally so in terms of marriage connections or distant ties of blood. People employed one another, or sold things to each other, or exchanged services, or simply rubbed shoulders here, there and everywhere. Each casual encounter in a lane or at a shop entrance reinforced this subtle and elaborate system, as hard to penetrate as any she had met. (pp. 71–2)

One day, while visiting the village shop, Stella listens to an interesting little homily on commitment from Molly the shopkeeper who is still unsure about Stella:

'You used to know how a person stood, without having to take soundings, know what I mean? You knew if they were farming or trade, church or chapel, you knew who their father was and which way they'd jump if it came to the push. Nowadays people can walk into the shop and it's anyone's guess, frankly . . .'

Stella then thinks to herself: 'I am part of the landscape like everyone else. And some of us are more tenuously placed within that landscape than others. Some are entrenched; others merely perch' (p. 175). The difficult challenge for Stella is to move from observer to participant. She never quite makes that transition, never quite fits in and eventually leaves.

Place and landscapes

Place involves 'a specific landscape, a set of social activities, and webs of meaning and rituals, all inseparably intertwined'.[34] Places are inherently associated with the events that happen in landscapes. Human memories, whether individual or collective, are so often localized in landscapes even when people cannot precisely remember when they happened in time or how long they lasted.[35] Landscape, then, is the first partner in the dialectical nature of place. Yet the very word 'landscape' implies an active human shaping rather than a pure habitat. Historically the land has been actively shaped for as long as humans have existed – by agriculture, forestation, enclosure, and, as far as England is concerned, aesthetic landscaping by the wealthy since the late seventeenth century. One only has to think how natural features, mixed with earlier human settlement and meanings, were given a new shape, a particular interpretation during the eighteenth century in the 'landscaping' of monastic sites in North Yorkshire such as Rievaulx and Fountains. This is artificial rather than natural. However, does 'artificial' ('by human artifice' or invention) necessarily imply inauthentic or inappropriate?

Simon Schama, in his monumental work *Landscape and Memory*, is clear that 'Landscapes are culture before they are nature; constructs of the imagination projected onto wood and water and rock.'[36] Our contemporary tendency to objectify, reify or even romanticize nature (whether plants or the animal kingdom) is born of anxiety. It reflects a human sense of distance from the natural world, a guilty recollection of abuse through unreflective use of industrial and technological power, and a subsequent concern to repair the damage through an appropriate ecological consciousness.

Apart from human embodiment the most common experience of place, or being placed, involves familiar landscapes. Any analysis of place inevitably has a subjective element. People learn to be who they are by relating to the foundational landscapes of childhood or to adopted landscapes that became significant because of later events and associations. Familiar landscapes are the geography of human imagination. Their power concerns more than beauty for landscapes are not necessarily inherently romantic or awe-inspiring. People cannot but be culturally conditioned in terms of the kind of landscape that exercises power over them.[37] It is interesting to me to ask what is particular about English landscape. One feature stands out. In such a small country that is relatively densely populated, there is very little true wilderness. In a book that many people consider groundbreaking, *The Making of the English Landscape*, W. G. Hoskins noted that there are not many places where one can feel with such complete assurance that this is exactly as the first inhabitants saw it in 'the freshness of the early world'. Not much of England, even in its more withdrawn, inhuman places, has escaped being altered by man in some subtle way or other, however untouched we may fancy it is at first sight.[38]

Hoskins himself admitted that everything is older than we first think. Yet he persisted in the view that a largely untouched natural landscape of primeval woodland was substantially altered only in early medieval times. Nowadays we know that forest clearance began as early as 8000 BCE and that what existed by the time of the eleventh-century land and ownership survey, known as Doomsday Book, was the result of successive phases of clearance, regeneration and new clearance. To put matters in simple terms, English landscape is archetypally a historical landscape.

My central point is that the power of landscape does not stem purely from something inherent in the topography or in some transcendent presence, the *genius loci* or the 'spirit of place' of Classical Mediterranean culture. If there is a given quality to particular places, this is partly topographical and partly a matter of the accumulation of human histories. Kathleen Norris in her best-selling essay on 'spiritual geography', *Dakota*, writes of 'the place where I've wrestled my story out of the circumstances of landscape and inheritance'.[39]

Although place is a human construct, it is equally vital not to lose sight of the fact that the natural features are part of the inter-relationships that go to make up place. The physical landscape is a partner, and an active rather than purely passive partner, in the conversation that creates the nature of a place. It is paradoxical that so much radical contemporary writing on the politics of place fails to mention the non-human element at all. This is simply to substitute a new anthropocentrism for old. However, a writer like Schama, for example, does not suggest that there is no real nature, merely that there is no *pure* nature. Rather, there is an interplay between physical geographies and geographies of the mind and spirit.

Place and memory

Schama is correct in reminding us that human memory regarding landscapes always has a more powerful effect on our thinking than the actual contours themselves. To put matters theologically, God is not revealed to us in the immediacy of raw nature. The only spirituality that is accessible is incarnational – that is mediated through the cultural and contextual overlays we inevitably bring to nature and to our understandings of the sacred. 'Every habitat is approached by means of a particular *habitus*, a way of reading the natural world that has accumulated over time.'[40]

If place is, therefore, first of all landscape, it is also memory. 'Places of memory. Places into which I had poured myself and all the longings of my life and which reflected back to me the shape and texture of my life there.'[41] Memory embedded in place, however, involves more than simply any one personal story. There are the wider and deeper narrative currents in a place that gather together all those who have ever lived there. Each person effectively reshapes a place by making his or her story a thread in the meaning of the place and also has to come to terms with the many layers of story that already exist in a given location.

My own childhood experience in the English county of Dorset was shaped by the fact that we were so visibly surrounded by old landscape marked by Bronze Age hill settlements, Celtic burial mounds or barrows, Roman military installations, medieval field systems, a myriad of medieval churches and monastic remains. Even the familiar clumps of ancient trees on local hilltops turned out to be related to sacred groves. There were mythic places too. Badbury Rings was the reputed site of *Mons Badonicus,* one of the legendary King Arthur's greatest victories. Badbury is an eerie and silent place. It was said in my childhood that no birds nested in the trees because of the great slaughter that took place there all those centuries ago. Collective memory materialized in landscape. It is difficult for one's identity not to be unconsciously marked by that sense of historical placement.

Landscape is not merely shaped by enclosure, agriculture, forestation or settlement. It is also named. Names give the landscape a particular meaning in relation to human memories. No name is arbitrary. Every name, even a single word, is a code that, once understood, unlocks a world of associations, events, people and their stories. Penelope Lively puts it rather well in the novel I have referred to (p. 94): 'Here was a relentless Anglo-Saxon plod, there was a faint Celtic whisper. Here a hint of Roman, there a Norman reference'. The ancient town of Salisbury where I live is an excellent example. The name is a Norman variant (substituting the 'l' for an 'r') of a Saxon name *Searobyrg* (derived from *Searaburg*) meaning 'armour fort'. But this in turn hints at the older Roman name *Sorviodunum*. The Saxons simply treated the first element of the Roman name as a sound to be transposed into their word for 'armour' and translated the second element into the Saxon equivalent for 'fort'. However, the Roman name itself has its older whispers for it is Romano-British and its second element *dunum* derives from the Celtic word *dunon*.[42]

It is appropriate to think of places as texts, layered with meaning. Every place has an excess of meaning beyond what can be seen or understood at any one time. This excess persistently overflows any attempt at a final definition. A place can never be subordinated to a single valuation, one person's prejudices, or the assumptions of a single group. The hermeneutic of place progressively reveals new meanings in a kind of conversation between topography, memory and the presence of particular people at any given moment. 'All human experience is narrative in the way we imaginatively reconstruct it . . . and every encounter of the sacred is rooted in a place, a socio-spatial context that is rich in myth and symbol'.[43] Hence, my fundamental contention is that there can be no sense of place without narrative.

Place and narrative

If place lends structure, context and vividness to narratives, it is stories, whether fictional or biographical, which give shape to place. However, as stories of displacement show, it is the absence of lineage and memory associated with physical place that is just as critical as separation from the landscape alone. The most important part of the family mythology of the Coloured actor Joseph Malan in André Brink's novel *Looking on Darkness* is that, unlike most Coloured and Black South Africans, they have a traceable history just like the Europeans.[44] In Brink's writings, landscapes and narratives are continually linked by his stark and painful question: Whose story is given space?

The French philosopher Paul Ricoeur, who has had a significant influence on theology, has been greatly preoccupied with the importance of narrative to human identity and with reconstructing a viable 'historical consciousness'. This, he argues, is vital to our individual and collective identities – and, implicitly, to our spiritual well-being: 'time becomes human time to the extent that it is organised after the manner of a narrative; narrative, in turn, is meaningful to the extent that it portrays the features of temporal existence'.[45]

At first glance, Ricoeur appears to be something of a paradox. On the one hand, he shares a postmodern scepticism with metanarratives and is profoundly suspicious of 'giving in to the temptation of the completed totality'.[46] Ricoeur shares with postmodernist thinkers the belief that we must renounce any attempt by history 'to decipher the supreme plot'. However, Ricoeur also rejects a tendency to equate this renunciation with the impossibility of history as a form of narrative. In fact he argues that the former search for a supreme plot or metanarrative actually undermined true narrative because it sought to transcend context and the particularity of all stories. As a result it reduced history to 'the totalisation of time in the eternal present'.[47] One might add, in theological terms, that the temptation of totalization too often made the Christian Church appear, without qualification, as a place of completed, comprehensive and exclusive meaning and explanation – fully realized eschatology all in itself.

Yet Christianity cannot ignore the fact that it does seek to speak a narrative of meaning, albeit one constrained by an internal tension between affirmation and apophatic denial. Riceour, too, recognizes that humans cannot live without narratives of meaning. If we reject the possibility of mediating narratives altogether this is not the liberating experience that it may appear. On the contrary, it is profoundly oppressive. The reason is that without narrative we risk two things. First of all we undermine a key element of human solidarity (we bond together by sharing stories) and second we are trapped in the immediacy of the present. We reduce or remove a key incentive for changing the status quo as well as an important means of bringing this about. 'We tell stories because in the last analysis human lives need and merit being narrated. The whole history of suffering cries out for vengeance and calls for narrative.'[48]

Narrative is a critical key to our identity, for we all need a story to live by in order to make sense of the otherwise unrelated events of life and to find a sense of dignity. It is only by enabling alternative stories to be heard that an elitist 'history' may be prised open to offer an entry point for the oppressed who have otherwise been excluded from the history of public places. 'Without a narrative, a person's life is merely a random sequence of unrelated events: birth and death are inscrutable, temporality is a terror and a burden, and suffering and loss remain mute and unintelligible.'[49] Rather than abolish narrative we need to ask, 'Whose narrative has been told?' 'Who belongs within the story of this place?'

Narrative and history are closely related. Ricoeur seeks to overcome the absolute dichotomy between history as 'true' and stories as 'fiction'.[50] For Ricoeur both history and fiction refer in different ways to the historicity of human existence. Both share a common narrative structure. Both employ a plot to suggest a pattern for otherwise episodic events. Any and every plot chooses a sequence for events and characters that suggests a direction or movement. This is shaped by a particular point of view.

Ricoeur rightly rejects the positivist myth of history that has prevailed since the nineteenth century (that is, history is only what is scientifically verifiable) in favour of one that allows for the presence of 'fiction'. That is, history is restored as a form of literature that does not simply recount events in a disconnected or disinterested way but *organizes* them in a form that seeks coherence. Equally, however, fiction may be *truthful* in that, while not slavishly tied to the mechanical details of events, it is capable of addressing something equally important about reality – the realms of possibility and of the 'universal'. History and fiction are both narratives that seek to describe either what reality *is* or *is-like* with the purpose of making human existence meaningful.

Ricoeur may be said to be attempting to retrieve history as something more than merely a disconnected set of cold, objectified 'events' emptied of the warmth of human stories. 'History' once again has something to do with people's vision. It is an act of interpretation and all interpretation is necessarily an act of *commitment*. History implies continuities and continuities in turn imply responsibilities. Commitment and responsibility point to the important fact that 'history' is not merely about the past but also about the present and future. A historical consciousness opens us to possible action rather than to a passive acceptance of 'the way things are'. In this way of understanding, 'history' becomes a critical spiritual and ethical issue.

Place and conflict

Ricoeur's concern to recover what we might call a 'narrative of the oppressed' returns us to my earlier comment concerning contemporary treatments of place by social scientists and anthropologists. These increasingly affirm that, because place is always a contested rather than a simple reality, the human engagement with place is a political issue. Place is also political because the way it is constructed means that it is occupied by some people's stories but not by others. Schama offers the fate of the Lithuanian forest, Bialowieza, and its primitive bison as a parable of the way landscape is always both an interpreted reality and at the same time a political issue. After the German invasion of 1941, the forest and its bison were the subject of visionary conservation by Reichsmarshal Herman Göring in his pursuit of Teutonic symbolism. The price was the eradication of all traces of Jewish and Polish peasant occupation. In order to create a protected forest zone, a 'total landscape plan' as it was called, the first task was to get rid of people and to empty the villages.[51]

By deconstructing modernity's belief in objective 'absolute' place, postmodern critiques affirm that definition is power. The

French Marxist philosopher Henri Lefebvre offers an analysis of place that also reminds us that systems of spatialization are historically conditioned.[52] Spatializations are not merely physical arrangements of things but also spatial patterns of social action and routine, and historical conceptions of the world. These add up to what Lefebvre calls a 'socio-spatial outlook' that manifests itself in our every intuition. The metanarratives of those with secular or religious power at any given time take over public places and thus become stories of dominance and repression. In the case of Christian theology and spirituality, as the Peruvian theologian Gustavo Gutiérrez reminds us, the Christian narrative was compromised by becoming synonymous with European culture and, for people of other cultures, with the values and experiences of colonizers from 'elsewhere'.[53]

We need to become aware of what might be called 'a longer narrative' in which 'the others' who have been made absent by those who control public or institutional histories are now being restored as people who are fully present. They are no longer a presumed and distant 'them' removed from a vague and tacit 'us'.[54] The notion that place relates to issues of empowerment and disempowerment forces us to think of multilocalities (that actual locations are many different 'places' at the same time) and multivocalities (that there are a variety of different voices to be heard in a place).

To be a person is not merely to be embodied but also to inhabit a *public* place. Our social selves are created for us, not just symbolically but also physically within roles determined by social, cultural and religious hierarchies as well as gender stereotypes. So, for example, we put on masculinity or femininity with clothes and manners in such a way that it changes the very shape our bodies occupy in space. Human places themselves can be read as landscapes of exclusion. So, the way people describe particular places as central or peripheral tends to accord with whether they are associated with high or low culture. Power is expressed in the monopolization of central places by socially strong groups and the relegation of weaker groups in society to less desirable environments.

ıcal reflections on place can no longer ignore that the
ɔncrete places is full of exiles, displaced peoples, diaspora
ties, increasingly inflamed border disputes and the violent
ɟ by indigenous people and cultural minorities to achieve
ʜᴇ— ɔn. While narrative is important, we have to reconstruct
what might be called a 'narrative beyond easy narrative' if space is
to be made for those whose stories have not been heard. A British
feminist theologian such as Elaine Graham faces squarely the
tension between reclaiming public place from a patriarchy that has
made women placeless and leaving such place behind in order to
create alternative territories. Like Ricoeur she suggests that all
that any of us have is spatial location – a historical context and a
material experience that is inherently contested. In other words,
there is no ideal place to withdraw to that is not political. A strategy
of resistance demands a critical engagement with contexts and
the political will to transform the narrative and to redeem the
place.[55]

Place and particularity

All this suggests to me that a theological attempt to reconstruct an
effective narrative of place, with room for the unheard or marginalized, must begin with a serious attention to the vocabulary of the
particular. The problem with the Western culture of 'modernity',
that has dominated our thinking over the last couple of hundred
years, is that its impulse is to stress the universal rather than the
particular or vernacular, the anonymous or disengaged rather than
the personal. The connection with tendencies in theology is obvious
although there are differing views about the degree to which it was
secular 'modernity' that influenced theology or tendencies in
Christian theology that gave birth to its secularized counterpart.
One might assume that a religion based on the doctrine of the
Incarnation would have been consistent in according a fundamental
importance to human history and to material existence. However, if
the stories of the origin of Christianity suggest an affirmation of
history, there has always been a siren voice that suggests that what
is fundamentally important lies in a spiritual and eternal realm on

the far side of time and place. I will explore this tension later, but at this point I would simply suggest that there could be no positive theology of place without an unequivocal engagement with particularity.

In seeking a theological vocabulary that enables engagement with particularity, it may be helpful to look again at one of the great figures of medieval philosophy and theology, the Scottish Franciscan Duns Scotus. Scotus flourished towards the end of the thirteenth century at Oxford, Paris and, according to some traditions, in Cambridge as well. Interpretations of Scotus, the *doctor subtilis*, have tended to be somewhat inconclusive and uncertain. This may partly explain why his theology has received less attention than that of Thomas Aquinas. However, while acknowledging the need for caution, it is generally agreed these days that Scotus' thought was expressed most originally in a theology of particularity and individuality.

At the heart of Scotus' high view of the particular lies his distinctive understanding of the Christian doctrine of Incarnation. Although Scotus is rarely quoted in books on spirituality, his philosophical and theological writings had an immense impact on the Franciscan tradition of spirituality for centuries and through this, albeit implicitly, on much else besides. For Scotus, the Incarnation is God's greatest work and cannot be explained by anything outside God's own reality and eternal intention, such as by chance or by human sin. Incarnation is 'the highest good in the whole of creation' and 'was immediately foreseen from all eternity by God as a good proximate to the end'.[56] By 'end' Scotus means God's purpose for creation. This purpose, for Scotus, is deification or a sharing in God's own life. God's life is so fruitful that it constantly and inherently seeks expression in the particularities of the created order.

Duns Scotus offers a theologically positive view of what is specific and individual, even the smallest of details. His theology of Incarnation leaves us with the thought that God not only creates all things *for* Christ but also in Christ. The nineteenth-century Jesuit poet Gerard Manley Hopkins was strongly influenced by Scotus. As Hopkins came to realize in more poetic terms, Scotus taught that everything without exception is rooted in the cause of creation. This cause is the humanity of Jesus. By implication all things exist not only to be themselves, and to 'do' themselves. Also, in this being and doing, all things 'do Christ'. Thus each individual or particular thing is more than simply a symbol of something more. That would make it dispensable – usable and disposable. One thing might be substituted for another if it proved to be a better symbol. There would thus be no unique value in any individual or particular thing. Here Scotus departs from the better known scholastic theory of analogy whereby true being exists only in God and everything else is derivative, pointing indirectly towards true being. Scotus, however, suggests that all things *in their very particularity* participate directly in the life of the Creator. This is one thing implied by Scotus' difficult concept of 'univocity of being'. Because everything participates directly in God, each thing is a uniquely important expression of God's beauty as a whole.

It is this perception that lies behind Hopkins' famous poem 'As kingfishers catch fire':

As kingfishers catch fire, dragonflies draw flame;
As tumbled over rim in roundy wells
Stones ring; like each tucked string tells, each hung bell's
Bow swung find tongue to fling out broad its name;
Each mortal thing does one thing and the same:
Deals out that being indoors each one dwells;
Selves – goes itself; *myself* it speaks and spells,
Crying *What I do is me: for that I came.*

I say more: the just man justices;
Keeps grace: that keeps all his goings graces;
Acts in God's eye what in God's eye he is –
Christ. For Christ plays in ten thousand places,
Lovely in limbs, and lovely in eyes not his
To the Father through the features of men's faces.[57]

As the poem 'Duns Scotus' Oxford' makes abundantly clear, Hopkins was ambivalent about the increasingly industrialized

Oxford of his day with its 'base and brickish skirt there'. But nevertheless his sense of place was more deeply charged not merely by a nostalgia for a lost 'rural rural keeping – folk, flocks, and flowers' but by a sense of the presence of Duns Scotus:

Yet ah! This air I gather and I release
He lived on; these weeds and waters, these walls are what
He haunted who of all men most sways my spirits to peace . . .[58]

This air, *these* weeds and waters, *these* walls . . . This is the key to what most attracted Hopkins to Duns Scotus and helped to shape Hopkins' extremely intense union with nature and in-depth sense of the doctrine of the Incarnation. Hopkins is, if you like, a poetic expression of Scotus' philosophy of specificity and particularity. This intense sense of God in each and every particular is what lies behind Hopkins' judgment that Scotus was 'Of realty the rarest-veined unraveller'.

Scotus' principle of individuation and his epistemology attached great importance to individuality and personality. For example, to the category of leaf or place is added an individualizing form, or final perfection, that makes this leaf *this* or this place *this* rather than that. Scotus gave the name *haecceitas*, 'thisness', to this individualizing form. Scotus believed that the individual anything or anybody is immediately knowable by the intellect in union with the senses. The first act of knowledge was therefore a recognition, albeit vague, of the individual, the concrete, the particular glimpse of 'thisness'. Only by knowledge of the particular and of the concrete are our minds able to arrive eventually, by a process of abstraction and comparison with what is experienced as 'like this, while not this', at the knowledge of the universal, 'leaf', 'place', 'person'. In Scotus' scholastic language: 'It is impossible to abstract universals from the singular without previous knowledge of the singular; for in this case the intellect would abstract without knowing from what it was abstracting' (*De Anima*, 22, 3). Incarnation, therefore, in Scotist theology takes on a special force.

Scotus raised 'the particular' from being merely an instance of something of a certain type, a mere exemplification of a category.

What is individual is just itself even if it is also related to similar but *other* realities. Thus for Scotus, what is particular and specific is more perfect because it is unique. Indeed, nature is predetermined to singularity.[59] *Haecceitas* is utterly specific and exclusive and is to be found only in this and that particular. This concept of absolute particularity, as opposed to the greater perfection of what transcends particularity or achieves a certain abstract universality, seems to me to accord somewhat with one aspect of contemporary postmodern sensibilities.[60]

It is interesting that Franciscan scholars nowadays emphasize that Duns Scotus' concept of the perfection of 'the particular' was influenced above all by the Canticle of Creation of St Francis of Assisi.[61] If this is the case, it is important to understand what the Canticle means. It is certainly possible to reduce it, as some more devotional approaches have done, to a rather bland, romantic love of the natural world. However, the underlying meaning of the Canticle is more complex than that. The key notion is that all our fellow creatures (whether animate or inanimate), as brothers and sisters, reflect to us the face of Christ. They 'do Christ'. Francis experienced each particular element of creation, not merely creation as an abstract whole, as coming from the same source, the good God of the Trinity revealed in the Incarnate Son. The corollary is that each created particularity is revelation. People may come to know God through each element. The foundation of Franciscan respect for all created things is that the One through whom everything was created has come among us to be a creature.

The first nine verses speak of the cosmic fraternity of all elements of creation.

Praise and glory, honour and blessing
Be yours, O Lord,
O Most High
O Most powerful.

Praise and glory, honour and blessing
Be yours, O Most High.
O my Lord, be praised.

Let everything you have made
Be a song of praise to you,
Above all, His Excellency the Sun (our brother);
Through him you flood our days with light.
He is so beautiful, so radiant, so splendid,
O Most High, he reminds us of you.

My Lord, be praised
Through our Sister the Moon and through each Star.
You made them so clear and precious and lovely
And set them in the heavens for all to see.
Through Brother Wind and Sister Water,
Through Brother Fire and Mother Earth,
Through Sister Death – be praised, O Lord,
Be praised.

The Moon and Stars are clear and dear and fair,
Through them be praised, and through the Clouds and Air
By which you nourish us each changing day;
Through precious Water, pure in every way,
So useful, humble, chaste, receive our song.

Through Brother Fire, robust and glad and strong,
None shines as he shines in blackest night,
How handsome he, how joyous and how bright.
O my Lord, be praised!

Let everything you have made
Be a song of praise to you,
Above all, our Sister, our Mother, Lady Earth
Who feeds and rules and guides us.
Through her you give us fruits and flowers
Rich with a million hues.
O my Lord, be praised.

This uplifting doctrine of cosmic fraternity, however, conceals a much sharper and prophetic implication. Francis does not simply celebrate God's goodness expressed in the world as God's gift. Verses 10–11 celebrate the peace that comes from mutual pardon or reconciliation. It is generally thought that the verses were written as part of a campaign to settle a dispute between the mayor and bishop of Assisi.

> Be praised, my Lord,
> Through those who forgive for your love,
> Through those who are weak,
> In pain, in struggle,
> Who endure with peace,
> For you will make them Kings and Queens,
> O Lord Most High.[62]

Thus the created world is a 'reconciled space' because of the fraternity of all things in Christ. There is no room for violence, contention or rejection of the 'other'. But there is also the question of what Francis understood by the 'other'. For this we have to recall one of Francis' foundational experiences. The 'other' for Francis had a very particular meaning. Behind the text of the Canticle, and underlying his whole theology of creation and Incarnation, is another text – the transformation of consciousness brought about by his early encounter with a leper. In the first three verses of *The Testament* dictated shortly before his death in 1226 Francis actually identified the first moment of his spiritual life with his encounter with the leper:

> 1. The Lord granted me, Brother Francis, to begin to do penance in this way: While I was in sin, it seemed very bitter to me to see lepers. 2. And the Lord Himself led me among them and I had mercy upon them. 3. And when I left them that which seemed bitter to me was changed into sweetness of soul and body; and afterwards I lingered a little and left the world.[63]

The meeting with the leper was not merely an encounter with human suffering but, in medieval terms, Francis was led to embrace the excluded 'other'. Lepers were not simply infected with a fearful disease. In medieval society they symbolized the dark side of existence onto which medieval people projected a variety of fears, suspicions and guilt that must be excluded from the community not merely of the physically healthy but also of the spiritually pure. Lepers were outcasts banished from society. They joined the criminals, the mad, the excommunicated and the Jews. In many respects, lepers were not only perceived as wretched and dangerous but also, because profoundly symbolic of the corrupt nature of flesh in general, as scandalous. Interestingly, their corrupted flesh was often associated in the popular imagination with the practice of illicit sexuality. There was more than a hint that the sickness was a divine punishment for corrupted sexuality or for heresy (which were often associated with each other in the medieval mind).[64]

Through the encounter with the leper, Francis came to see that both participation in human experiences of suffering and of exclusion were at the heart of God's incarnation as revealed in the face of the Crucified Christ. If Duns Scotus' theology of the particularity of creation is a kind of exposition of the spirituality of the Canticle of St Francis, then *haecceitas*, 'thisness', necessarily involved a sense of God's place among the rejected and the garbage of this world. 'Thisness' expresses the absolute inclusivity of God that draws all things together within difference. I wish to return in more detail to a theology of reconciliation and inclusion, under the label of 'catholicity' (as a vital human value) in a later chapter.

Meanwhile, the image of the leper becomes for both St Francis and Duns Scotus a paradigm for our understanding of creation, Incarnation and thus discipleship. By entering the world of the concrete, specific and particular, and by taking on our flesh, God in Jesus becomes committed to, and thus redeems, *all* that humanity is, including what is unacceptable and 'other' and all places where humans dwell. Heidegger's concept of 'dwelling' suggests commitment. 'And the Word became flesh and *lived* among us' (John 1.14). The Christian doctrine of Incarnation offers an image of God's irrevocable commitment as *remaining*. Similarly, it seems to me, Duns Scotus' very Franciscan concept of *haecceitas* demands that believers become similarly engaged with particularity, with contingent reality, with specific places. By remaining *here* and *there*, Scotus proclaims, we encounter the face of God.

Departing from the particular

However, in Christian terms, a theology of place must maintain a balance between God's revelation in the particular and a sense that God's place ultimately escapes the boundaries of the localized. As we shall see in later chapters, there is a persistent tension in Christianity between what is sometimes referred to as place and placelessness or, as I prefer, between the local and universal dimensions of place. As a matter of fact, in general, intense experiences of place often provide people with their first inchoate intimations of transcendence or the sacred. Place is both *this, here and now*, and at the same time more than 'this', a pointer to 'elsewhere'.

In the light of Incarnation, spirituality is undoubtedly concerned with how to live within the complex world of events. Our place is specified by God's commitment to the particularities of the world and of human history. The event of Jesus Christ is set in a particular time and place. Yet there is a tension that is expressed somewhat elusively by Michel de Certeau:

> Christianity implies a *relationship to the event* which inaugurated it: Jesus Christ. It has had a series of intellectual and historical social forms which have had two apparently contradictory characteristics: the will to be *faithful* to the inaugural event: the necessity of being *different* from these beginnings.[65]

In de Certeau's terms, the particularity of the event of Jesus Christ is the measure of all authentic forms of Christian discipleship in the sense that they presuppose that event but are not identical repetitions. In a sense, the particularity of the event of Jesus Christ 'permits' the placed nature, the particularities of all subsequent discipleship. There, too, God may eternally say 'yes' to us without condition. The world of particular places is therefore the theatre of conversion, transformation and redemption. However, Jesus' place is also marked by an empty tomb. 'He is not here; for he has been raised, as he said . . . indeed he is going ahead of you to Galilee' (Matt. 28.6–7). God in Jesus cannot be simply pinned down to any here and there, this and that. The place of Jesus is now perpetually elusive. He is always the one who has gone before. To be in the place of Jesus, therefore, is literally to be disciples, to be those who 'follow after' in the direction of Jesus' perpetual departure.

It may be true that an emphasis on particularity alone merely cloaks fragmentation, incoherence or the refusal of narrative movement. However, it is equally important not to see the movement to transcend particular and local place as a denial of the world or of the value of human history. Rather it points, first of all, to the 'catholic' sense that the divine presence cannot be imprisoned in any contracted place or series of places. The divine is to be sought throughout the *oikumene*, the whole inhabited world (or, indeed, eventually the *oikumene* of the cosmos whatever that may ultimately mean). Again, to adopt the words of de Certeau, discipleship simultaneously demands a place and an 'elsewhere', 'further', 'more'. I believe that it is impossible to grasp the heart of de Certeau's perspective without explicit, even detailed, attention to his Jesuit roots and life-long preoccupation with the Ignatian mysticism of 'practice' or 'action' especially through the medium of the life and writings of Joseph Surin. This mysticism of practice offered de Certeau among other things the language of the *magis*, the *semper maior*, the always greater, the always more, the always beyond, together with the values of movement or trangressing boundaries, always exceeding limit in search of *oikumene*:

Within the Christian experience, the boundary or limit is a place for the action which ensures the step from a particular situation to a progress (opening a future and creating a new past), from a being 'there' to a being 'elsewhere', from one stage to another . . . A particular place – our present place – is required if there is to be a departure. Both elements, the place and the departure, are interrelated, because it is the withdrawal from a place that allows one to recognise the enclosure implicit in the initial position, and as a result it is this limited field which makes possible a further investigation. Boundaries are the place of the Christian work, and their displacements are the result of this work.

And again:

> In order to pass from one place to another, something must be
> *done* (not only *said*) that affects the boundary: namely, *praxis*. It is
> this action which transcends, whereas speeches and institutions
> circumscribe each place successively occupied.[66]

The catholicity of place is, for Christians, symbolized most
explicitly in the *koinonia* of believers filled with the Spirit of Jesus
and shaped by the space of the Eucharist. As we shall see later,
eucharistic space has a particular potency in terms of the tension
between local and universal. On the one hand, every Eucharist
exists in a particular time and place. On the other hand, each
Eucharist is both the practice of transgression and a *transitus* or
transit point, a passageway between worlds that prefigures the
conclusive 'passing over' that is ultimately brought about in death.
eucharistic space enables the particularity of local place to intersect
in the risen and ascended Jesus with all time and all place.

2

Place in Christian Tradition

Before reflecting on place in a more intensively theological way, I want to address a number of historical questions. Given the importance of place in human culture, as well as the centrality of land and temple in the theologies of the Hebrew Scriptures, it is strange at first sight that the Christian tradition as a whole makes little direct reference to place.

The apostolic period

Christianity came of age in the Jewish urban Diaspora of the Graeco-Roman world. The particularity of place could not be bypassed entirely given the fundamental nature of the doctrine of the Incarnation. It is expressed implicitly in the diversity of the Gospels, written with self-conscious attention to distinctive localized audiences, in the Pauline letters targeted at communities named after the places they inhabited and in the specific judgments pronounced on different local churches in the book of Revelation. However, it is also clear from these foundational texts that there was a different and urgent concern for Christian disciples in the apostolic era. This was to move out from what was local to them, from 'home', into the entire inhabited world, the *oikumene*, in advance of the rapidly approaching last days. Indeed, Acts 1.8 suggests that Jesus explicitly exhorted his disciples to move beyond the city of Jerusalem to the ends of the earth in pursuit of their mission to preach the Kingdom of God. For Christians, God was increasingly to be worshipped in whatever place they found themselves. The first martyr, Stephen, is described as dying precisely for questioning the supreme sacredness of the Temple and for reminding people that God 'does not dwell in houses made with human hands' (Acts 7.48). The experience of 'being in transit', of journey, became a central metaphor for encounter with and response to God. For the disciples, significant conversion experiences often occur 'on the way' in situations of displacement or transition, for example the disciples on the road to Emmaus (Luke 24) or the conversion of Saul on the road to Damascus (Acts 9). In a sense, it seems that the marginal ground *between* fixed places is where God is most often encountered.

Many New Testament texts also reflect a strongly apocalyptic perspective. The localized place of the Jewish 'Promised Land' becomes largely symbolic. The focus of the earliest Christian groups appears to have been on the impending overthrow of all earthly places and the imminent establishment of the Kingdom of God. So, Christianity came into being in the context of intense future-oriented expectations. The power of the gospel for Christians lay in a belief that the 'event' of Jesus Christ was God's definitive act.

Augustine's theology of history

From the end of the apostolic period, as immediate apocalyptic expectations receded, an inherent tension in the Christian world-view emerged more clearly. Is meaning to be found in 'the place of the world', the 'place of human events'? Or is meaning *only* to be found in some 'elsewhere'? One of the most powerful expressions of this theological tension in the patristic period lies in the theology of Augustine. Augustine's theology of history, expounded mainly in his *City of God*,[1] has probably been the single most influential historical theory in Western Christianity. Augustine's thought has also been pervasive in a more implicit way in Western culture in general. The particular context for the work was the sack of Rome by the Visigoths in 410 CE. This event critically undermined a perception, current among many Christians after the conversion of the emperor Constantine, that the Roman Empire was a divinely willed, and even divinely guided, framework for human existence. There had been a shift from a pre-Constantinian conflict between

Christianity and public history to the adoption of the faith by the political classes of the Empire. There can be little doubt that the rapid expansion of monasticism in the fourth century CE was partly a prophetic, spiritual and social reaction to the perception that the Church was becoming over-identified with public Imperial history.

In the context of a fragmentation of Empire in the first part of the fifth century CE, Augustine redefined the relationship between the Kingdom of God and human history in terms of two communities or forces, the City of God and the Earthly City. Like the tares and the wheat of the gospel parable, the two cities cannot be perfectly differentiated within time. Ultimate meaning and stability cannot be found in the Earthly City, which is a context of incessant change. Nevertheless, states and empires may be said to have a 'historical role' in that they have an impact on the advance of the City of God. Thus, the Roman Empire may have had a 'mission' to unify people in order that the gospel might spread more easily. However, in itself the Empire had no eternal significance and its historical decline was inevitable.

Augustine's distinction between the City of God and the Earthly City threw into sharp relief a tension at the heart of Christian spirituality. If the origins of Christianity involve an affirmation of history, there has always been a siren voice that suggests that what is fundamentally important spiritually lies *now* in a parallel dimension alongside or outside historical events and *ultimately* in an eternity on the far side of time. Augustinian theology and spirituality may be a reaction against a dangerous and theologically dubious association between God and the state. But is there a place in Augustinian thought for the Kingdom of God to grow *within* human history? Does the narrative of human places have meaning?

Because Augustine believed that the full explanation of 'history' lay in the ultimate 'event' (the consummation of God's Kingdom), his history belongs in the eschatological category. This concept of the divine control of history continued among Christian writers in a fairly uninterrupted way until relatively modern times. In the course of the last two hundred years or so this became absorbed into a post-Enlightenment belief in 'progress' – that the world was inexorably moving towards greater rationality, greater justice, greater civilization and greater economic development. Augustine himself must be exonerated from such confusions. He did not identify divine control of history with the concept of social, political or economic success. No historic age grew closer to God, perfection or eternity than any other did. In that sense, the City of God operates in a realm distinct from everyday history. Conversely, a providential understanding of history does not involve a triumphalistic theology but a theology of hope glimpsed through tragedies and failure.[2] This is just as well, because the Enlightenment interpretation of history as the triumph of progress died during the twentieth century somewhere between Paschendaele and Auschwitz.

An important key to Augustine's theology of history is that his *City of God* is based upon his own experience. Indeed some see it as an application on a wider canvas of Augustine's sense of God's providence in his own life as expressed in the pages of the *Confessions*.[3] Fundamentally, the lesson both of Augustine's chequered career and of world history is that out of all things comes good. Although the City of God operates on a level distinct from the events of ordinary history, the distancing is far from absolute. True, Rome (or any other human *imperium*) is not the Kingdom of God in human form. The Earthly City, and thus human history, is always contingent. But human history is God's creation and is not, therefore, to be condemned as merely evil, or treated as an illusion. For Augustine, the lessons of history lead him to be calmly confident about the future even *in and through* the ambiguities and darkness of human events and experiences. If contingent 'history' and 'the world of places' are to end, it is an ending that speaks of fulfilment rather than of destruction and therefore of ultimate meaninglessness.

Even if Augustine rejected a progress model of human history, and no age could be said to be closer to God than any other, he also possessed a deep sense that the world of places and *each and every moment* was equally filled with God's presence and activity. Augustine's theology of history essentially describes two dimensions of one history: the history of contingent events in the world of particular places and, running through it, the thread of sacred history that alone tells us what God is *really* doing.[4] Therefore, every moment is significant even if that significance is presently mysterious. Such a view seems close to Paul Ricoeur's understanding of 'narrative' which is not descriptive of the time–space world as it is, and therefore of 'history' in a positivist sense. Narrative *re*describes the world rather than describes it. Narrative brings together and harmonizes the otherwise discordant and disparate elements of the experience of time and history.

In an important sense, Augustine's distinction between sacred and secular history does not render the history of time and places meaningless. What Augustine rejects is the *ultimacy* of the contingent world of particular places. This, it seems to me, liberates time and place in their contingency from any need to be tidied up. It is only when a providential model of history collapses into an Enlightenment version of 'history-as-progress' that we end up with a re-editing of human history that becomes exclusive, sanitized, oppressive and dysfunctional.

People as the place of the sacred

The Christian community also developed from its earliest days a concern with the ways in which the *people* of God are the place of God. In the first place, Christianity was powered by a belief that revelation was focused not on a land or a temple, but on a person, Jesus Christ. Although the traditional Jewish sacred places continued to have some importance and appeal, this was primarily because they were places where Jesus, the source of meaning and the focus of hope, lived, died and was resurrected. So place became a spatial expression of a life, a teaching and a theology. The itinerary of pilgrimage to the 'Holy Places', for example, is governed by the texts of the Gospels. What matters is not the places themselves (some have moved over time) but what happened in them and how, in a quasi-sacramental way, the believer may be brought into contact with the saving events.

A fundamental locus of 'the holy' was also the community of believers who in baptism were gifted with the Holy Spirit.[5] This is certainly a significant emphasis in the Pauline corpus. The primary sacred place is, singly or collectively, described as the temple of the Spirit in women and men of faith. The locus of the sacred transgresses former boundaries and is to be found particularly where people seek to be a community in Christ, distinguished by the destruction of traditional separations and by a quality of common life. Christ is to be met in those who are rejected or excluded in the old religious and cultural dispensation – the hungry, thirsty, naked, imprisoned, women, Samaritans, Romans, publicans, sinners (Matt. 1.23; 25.31ff.).

In a more general way, the developing Christian tradition substituted the holiness of people for the holiness of sacred places. Places could be said to be sacred by association with human holiness. The rise to prominence of the Christian Church throughout the Mediterranean world between the late third and the mid-fourth centuries caused the location of the sacred to shift. The *locus* of supernatural power was increasingly focused on a limited number of exceptional human beings. Thus the Christian Church, in Peter Brown's pithy phrase 'the impresario of a wider change', produced increasing numbers of holy men and women who were tangible links between heaven and earth. The primary examples were the Apostles – and the places where they were supposed to have worked, died and been buried. However, there were also the martyrs of the second-century and third-century persecutions as well as some theologically formidable early-fourth-century bishops. Finally there were successive generations of holy men and women who, as hermits or in new monastic settlements, ringed the cities of the Christianized world from the days of St Anthony of Egypt onwards.[6] Holy people and their stories, more than any other medium, localized the Christian God. In acts of healing, of intercession with God or with human power structures, not to mention in offering good advice, holy men and women from late antiquity to the Middle Ages and beyond personalized the forces of the cosmos for ordinary believers.[7] If holy people were portrayed as associated with the extraordinary, the beyond and the Other, it was very important that they also remained highly visible carriers of spiritual traditions and values that, through baptism, were the new birthright of ordinary believers. However, because holy people were a kind of 'place of otherness' in the midst of the community, they were frequently angular and awkward as well as useful.

The result of focusing the sacred on people was shown especially in the physical contours and piety of the Christian Church. Among the abiding monuments to the sacred were the 'living saints' who gathered in organized communities. Thus, one of the earliest recorded Christian pilgrimages to Palestine was specifically to visit the *monks* as much as the Holy Places. The particular late-fourth-century pilgrim in question, Egeria, kept a record of her travels. She probably came from Spain and obviously belonged to some form of community of dedicated women for which her endearing account was written. She recorded, 'We had the un-expected pleasure of seeing there the holy and truly dedicated monks of Mesopotamia, including some of whose reputation and holy life we had heard long before we got there. I certainly never thought I would actually see them.'[8]

The celibate ascetics manifested graphically to the community that the era of the Apostles and the martyrs was not over and so they became star attractions in themselves. They incarnated the power of faith to control nature and to witness to the presence of God. It is recorded that a young pilgrim named Daniel set out for the Holy Land but reached no further than the eccentric ascetic Simeon Stylites living on his pillar near Antioch. It seems that this phenomenon was sufficiently striking for Daniel to remain and eventually to take up the same lifestyle himself. People were fascinated with the prodigious behaviour of the living saints. This obsessive attention to holy activities and behaviour explains why we still know that Simeon Stylites had a practice of touching his toes 1244 times in bowing before God![9] Interestingly, Simeon built his pillar alongside a major Roman highway. This ambiguity of location that combined solitude with a certain accessibility to visitors became characteristic of generations of Christian ascetic communities and individuals. We can see this nearer to home in the places chosen by early medieval Celtic ascetics. In Ireland, for example, the lowland countryside was predominantly a water-logged landscape with no roads and few tracks. In such a context, the hilltops, wild headlands and rocky islands where hermits and communities chose to live were not simply places of separation but were also often places of connection.[10]

Because of the emphasis on people as *loci* of the sacred, it is not surprising that a genre of literary portraiture, known as hagio-graphy, was born. In these tales, from late antiquity to the Middle Ages, religious and spiritual values were summed up and made effective through the medium of outstanding individuals. Hagio-graphy was less concerned with historical accuracy than with confirming a particular pattern of behaviour. Or, to put it in more technical terms, while modern scholarship concerns itself with 'the world behind the text', the authors of hagiographic narratives were concerned with the world *in front of the text*. In other words, the crucial question was what the story offered to readers that might enable them to live better lives or be more faithful Christian disciples.[11]

Modern historians subordinate their accounts to chronology and tie events to particular contexts. Hagiographers are largely indiffer-ent to these requirements. The time and place of a saint are in some sense quite incidental. Saints tend to be portrayed as having wisdom regardless of their human experience and age. Equally, because the saint abides outside the normal constraints of time and place, belonging essentially to a world of eternal truth, the earthly historical world is purely transitional. Thus hagiography employs such words as 'captivity', or 'pilgrimage in a foreign land'.[12]

As we have already seen, Paul Ricoeur seeks to overcome the absolute dichotomy between history as 'true' and stories as 'fiction'.[13] For Ricoeur both history and fiction refer in different ways to the historicity of human existence. Both share a common narrative structure. Both employ a plot to suggest a pattern for otherwise episodic events. Fiction may be *truthful* in that, while not slavishly tied to the mechanical details of events, it is capable of addressing something equally important about reality. In Riceour's sense, hagiography is a 'fictive' narrative that describes an imaginative world that transgresses the constraints of 'what really happened' in the positivist sense in order to give expression to what ideally ought to have happened and thus, by implication, to the promise of what *may* happen.[14] In a certain way, therefore, the tight conventions of such narratives do not make the story they are concerned to tell less true. What is of interest, however, is the saint himself or herself as an embodied location of human aspirations and expectations.

Hagiography can serve as a form of historical deviation that reinforces the status quo. There can be no doubt that the genre has sometimes operated in this way.[15] However, these stories also serve a more positive purpose. They are a form of narrative that operates on the level of myth. The tale is deliberately idealized, because its purpose is not to illuminate questions of concrete context but to sketch a kind of utopian place in which everything is properly ordered in terms of ultimate meaning. Hagiography as a literary form is quite explicitly a rhetorical device to capture the attention and ultimately the loyalty of the reader to a particular world of meaning. Hagiography speaks of the realm of possibility. In other words, the saint as *locus* of the sacred is the anticipation in the here and now of the ultimate shape of the Kingdom of God in which all is harmonious and reconciled. Hagiography was the celebration of the irruption of godly or paradise place into the space of the ordinary. The medium for this irruption was a departure from civilized place and social norms to a liminal state. Here, sacred order was paradoxically encountered in the place of wilderness, wildness and danger. 'Hagiography . . . was considered to be ethically rather than factually true. To move men [*sic*] to the love of God was such a noble purpose that other considerations (such as truthfulness) paled before it.'[16]

Although the change of saintly scene was associated with leaving home in the spirit of Matt. 19.29, there was also a significant move in terms of spatial codes. The desert (in a literal sense or wilderness more generally) involved a move away from a social centre to a lonely margin. The saint reminds people that, for all that human existence is social, humans in the end have to travel alone across the ultimate divide between this world and the other world. The ascetic holy person exemplifies this by leaving behind all social bonds and becoming a stranger.[17] The ascetic's location, geographically and morally, was defined by God not by family or social relations.

As we shall see in more detail in Chapter 4, the theme of utopian paradise also makes a strong appearance in early monastic texts. In moving away from the *polis* in favour of the *eremos*, the ascetic wanderer was not in some simple way rejecting culture for nature – after all, the desert signified wildness, danger and suffering rather than beauty or romantic harmony. Rather, the monastic ascetic sought a third way – a reconfiguration of disordered place into a restoration of a prelapsarian paradise that was, at the same time, an anticipation of the final restoration hoped for in the coming Kingdom. As we shall see, one of the strongest symbols of this restoration and anticipation was the ability for humans and wild animals to live at peace with each other within the utopic space of monastic enclosure.

One of the key motifs in the concept of the holy person as *locus* of the sacred is the passionate search for the always *more*. The theme of the 'more' is present in spiritual literature from patristic times onwards. Thus Abba Paphnutius is recorded as telling a tale to fellow monks of his journey to seek those who were truly living the gospel. 'He said that I should flee to the further desert so I could see whether there were any brother monks in the farthest reaches of the desert living as servants of Christ Jesus.'[18] 'Farthest' is equated with 'better' or 'more', and refers both to 'further away from' the normal and to something that was perpetually receding and therefore elusive. Saints are described in terms of those who are perpetually dissatisfied with where they are. They are driven onwards always to seek a better way of life, or a person holier than them. In the case of Abba Paphnutius, before he finally found Abba Onnophrius, he had to cross a number of thresholds representing possible satisfaction. However, one potential teacher is found dead and another will not allow him to stay.

The theme of 'the always more' has an extensive life. It is picked up once again in another form in the sixteenth-century spirituality of Ignatius Loyola, particularly as expressed in *The Spiritual Exercises*. The text, of course, centres on discipleship – the call of Christ and how we may become ever more free to respond wholeheartedly. At a number of critical points when Loyola discusses discernment and making choices, the main criterion is described as 'the more'. What is *more* conducive to our purpose as human beings, what points towards *greater* praise of God?[19] This *magis* or *lo mas* is associated with generosity and freedom. The term is also used at other points in the *Exercises*. For example, there are two variants in the Meditation on The Three Classes of Persons. The stated purpose is 'to aid one towards embracing what is better' (Exx 149) and the prayer is that we should 'desire and know what will be more pleasing to the Divine Goodness' (Exx 151).

Some holy people existed marginally in the midst of or alongside the community. Others expressed their marginality by living at a site 'associated, whenever possible, with the wild antithesis of the settled land'.[20] The pilgrim not only had to leave the familiar but to arrive at a place that was self-evidently other than the everyday. What gradually established the ascetics as *loci* of 'the holy' was not, in the end, what I have already mentioned – that is, their healing or other useful powers, ability to predict or to offer discerning advice about life, the universe or merely the trivial. Much of this was pretty familiar stuff in the context of an extraordinary range of eclectic spiritualities on offer in the late Roman Empire. No, what was specifically Christian was that all these things evoked and imaged the deepest elements of faith and practice. The point was not that the ascetic was trying to be extraordinary – an ascetic or a miracle worker – but trying to be a *disciple of Christ* like everyone else. However, discipleship in the form of the holy person was manifested with a kind of intensification that concentrated in a small, embodied space and particular geographical location what would normally be diffused and dispersed throughout the community at large. This space for the holy was utopian in the sense that it expressed the desires and hopes of the general Christian multitude.

An important element of this intensification was the miracle – an unusual phenomenon that violated the usual course of nature. What is interesting is that the concept of miracle unites, just for a moment, two places, two worlds. It occurs in ordinary time and space but the power is a manifestation of other-worldly place. The miracle overcomes the everyday dissociation of two worlds and reveals their intimate not accidental connection. A miracle, therefore, explains God's world as an integrity because it reveals it all at once whereas in everyday life the eternal and the mundane are experienced as in opposition. While in patristic and medieval theological writing, the earthly city and the heavenly city were critically separated, in popular literature about saints and miracles they are portrayed as extremely close, in constant contact and association.[21] The saint belongs at once to both worlds because already while alive he or she is a citizen not only of the *polis* but also of the heavenly city. Because the two worlds are so interwoven, their laws are difficult to distinguish particularly when encapsulated in the actions and behaviours of holy people.

The problem with local saints

The change of scene central to the stories of saints is also associated with the fact that relatively few saints, until the later Middle Ages, seem to have been able to remain at home and attain sanctity. Perhaps, more accurately, they found it hard to be accepted as a saint without upsetting the status quo. They often experienced for themselves the force of Jesus' words 'Truly I tell you, no prophet is accepted in the prophet's hometown' (Luke 4.24). This had less to do with incredulity at the wondrous behaviour of the saints as with their role. There were some exceptions. By the thirteenth century, the development of Italian communes became associated with the desire to find a local holy person as the ultimate endorsement of their civic virtue as well as worldly success. A native holy person could sanctify the very space of the city and endow its history with religious authority. Interestingly, the candidates chosen were frequently not the great and the good but often humble penitents whose role was not merely to expiate their own personal sins but to act in an expiatory role for the whole commune.[22]

More generally, the fourteenth century onwards saw a change in the social significance of the cult of saints. In practice the promotion of human holiness moved from being largely a question of fulfilling the interests of social or religious elites to being a more popular creation. Hardly surprisingly, lay people preferred intercessors who were closer in space (local) and time (relatively recent) than had hitherto been the case.[23] It is also notable that during the Reformation the most violent popular resentment against the iconography of saints and their shrines was expressed in northern Germany and in Scandinavia. This cannot be entirely explained by the power of state officials (which applied in England with much less effect) or by the power of Reformation theology. It seems that the riots were fuelled by xenophobic motives for the simple reason that few 'local' saints in those countries were in fact local at all. Many were imports at the behest of powerful prelates or other social elites and the promotion of their cults was interpreted as the exploitation of the honest locals.[24] The contrast with the numbers of local saints in England and Wales is striking as is the maintenance of traditions of pilgrimage on the part of local Protestants as well as Roman Catholic recusants to shrines theoretically destroyed at the Reformation, such as St Winifred's Well at Holywell.

At the same time it is possible to detect some evidence for the localized quality of holiness. Saints and perceptions of holiness were closely entwined with a particular community that above all had a territorial dimension. Place not only influenced the progress of saintly careers but also set spatial limits around a cult. Taken as a whole, the genre of saints' lives sharply distinguishes Northern Europe from the Mediterranean countries, at least from the thirteenth century. Thus, rather crudely, Tuscan hill towns offered a setting for the conversion of adolescent girls and for miracle workers while Rhinelanders were more inclined to venerate great bishops or theological mystics. The supernatural activity of holy people within what is now modern France reflects the fact that this area was a cultural and geographical crossroads. Holy men and women range from powerful princes (similar to the Rhineland) to the miracle workers similar to Tuscany. The Mediterranean saints seem more likely to have come from non-aristocratic classes and to have been involved in some kind of family feud. They are also more likely to have had problems with sexual temptation. In terms of popular piety, therefore, Europe was substantially polarized centuries before the Reformation. If we are to believe the French historian Fernand Braudel, Europe divides along the line of the olive trees. His distinction seems to have some validity in reference to the complexities of religious symbols and spiritual behaviour as well as in reference to the economy![25]

The holy person as *locus* of the sacred had an ambiguous quality. Clearly each saint was particular and in that sense was embodied as a 'local' place. Yet hagiographically, saints and ascetics conformed to general types and these types reflected values that, while culturally variable to some degree, were also treated as universal Christian images of the perfected person. In a sense, the holy person as sacred place held together the same tension that we have already noted in a more general Christian approach to place – that is, between particularity and universality. Another ambiguity is that holy people both reflect what is ideally valued and at the same time subvert what is existentially practised. They are also symbols of exclusion and inclusion. They exclude what is interpreted as fundamentally destructive, sinful or conventional in the world's eyes, yet their tales (like the story of St Francis and the leper, and many other similar narratives) regularly make them a reconciling space that includes much that is 'other', dangerous and marginal to the prevailing order of things.

The canonization processes, or the local cultus of saints, have frequently reflected the values of the 'establishment', whether secular or religious, and sanctioned certain ideals, or official reforms or the respectability of particular religious communities. As we have noted, the recognition of the holiness of one of its citizens could also enhance the reputation of a particular city. Apart from 'establishment' saints, however, there were many popular saints, not least in the Middle Ages, some of whom were never canonized or had their popular cultus confirmed by the institutional Church. These popular saints seem to manifest two characteristic impulses of general piety. First, they were a kind of ethical space that preserved the ideal of moral purity. Second, while existing in a particular place and time, they pointed to the possibility of transgressing normal constraints by achieving some kind of entry into God's place. Saints were a form of subversive space because they stood as a reproach to the materialistic values of society at large. They were also a form of utopian space in that they incarnated the values and virtues that people aspired to, even if their achieval was beyond the powers of ordinary mortals. In recent years there have been a number of historical studies of saints from the point of view of social history. These ask what we can learn about the values of particular societies from a study of how saints and holiness were viewed. Conversely, such studies suggest that understandings of holiness mirror the social values of particular times and places, or the conscious reversal of such values, and are not completely autonomous categories of experience.[26]

I have written elsewhere about some of the ways in which saints have been used to sanction official structures.[27] Here I want to concentrate on how holy people, and stories about them, also subverted conventional boundaries to create an alternative space. Francis of Assisi's choice of radical poverty was a rejection of what he understood to be the characteristic sins of his time and his social class. So, we would be unhistorical if we accepted in an uncritical way St Bonaventure's portrayal of Francis' decision to lead a life of poverty as simply the result of hearing at Mass the reading from Matt. 10.[28] His phrase in the *Later Rule* 'as pilgrims and strangers in this world who serve the Lord in poverty and humility, let them go begging for alms with full trust' expresses an understanding of discipleship that accords substantially with the evangelical movements of his day.[29]

In fact between 1000 and 1700 approximately 75 per cent of saints belonged to socially elite classes. There was a tendency to equate moral with social nobility. On the other hand there are two examples of the holy person as a subversive figure. In the first place, holiness was conceived in terms of a reversal, especially of worldly status. In many respects, the whole tenor of the great monastic rules (which we shall examine in more detail in Chapter 4) pointed to a world in which conventional status was reversed. However, in stories of saints, the surrender by a wealthy or noble saint of position and riches was more powerful than the nothingness of the already poor. Christianity, in ways similar to the myths of other world religions such as the story of the Buddha, was particularly struck by a spiritual journey that involved a story of material riches to rags and then back to spiritual wealth. Second, there was a tradition of servant saints which subverted the conventional social order. The usual story involved a woman, normally living in servitude and badly treated in a variety of ways. The behaviour of the servant saint acted as a striking critique of the brutality and abusive behaviour of the (usually male) master. In the first place, the servant saint was portrayed as achieving a truly Christian life of virtue and noble human life despite the dual handicaps of social nothingness and of brutal abuse. Often the servant saint, for example St Gunthild of Suffersheim (mid-eleventh century) also acted out the true nature of nobility by her works of charity to the poor whom her supposed noble master had defrauded.[30]

Burials and shrines

Cemeteries were sacred from the earliest times. Holy men and women eventually died. Thus it was their burial places, along with those of Apostles and martyrs, which became the most visible monuments of this changed understanding of the sacred as located in people. It was important that the holy dead should continue to exercise their drawing power in the midst of the living. Because a theology of resurrection altered the meaning of death to point onwards to another form of existence, dead people had a special role in Christianity by joining two worlds together. Their tombs were privileged places where contrasting worlds could meet.

The traditional practice in the pre-Christian Roman Empire was to relegate cemeteries to outside the walls of cities. By the end of the sixth century, the tombs of saints were becoming centres of public Church life. There was a kind of theology of real presence that suggested an equation: the saints are 'with God'; the saints continue to be with us; ergo they are mediators in our midst of the presence of divine power – a kind of 'taster' of what was on offer in the other world. This produced a reversal of traditional beliefs and practices about death and burial and was one of the most powerful symbols of the encroachment of a specifically Christian culture into the mainstream of the late Empire. The dead were no longer rigidly excluded from the public life of the city of the living. From the start of its public existence, Christianity engineered a massive and subversive transgression of important boundaries.[31]

All this was reflected in the growth of Christian architecture. The monastery of St Simeon Stylites built towards the end of the fifth century had one of the largest churches in the eastern Mediterranean to mark the spot of the saint's various progressively higher pillars. The ruins are still visible. In a certain sense as well, shrines did refocus the sacred on place when Christianity moved its centre of gravity from missionary journeys to settled urban communities around the Roman Empire. The great civic buildings of Roman cities were mimicked in the cemeteries in order to accommodate the liturgies of pilgrimage. The monastic leader Paulinus of Nola actually boasted of creating around the tomb of St Felix a complex that could easily pass for a town.[32]

The relationship between this shift of sensibilities and the development of monasteries is interesting. The connections between early monastic theory and the continuation of martyrdom spirituality are well established. However, what is of concern here is that Christian pilgrims began to patronize not only the dead but also those who might be thought of as the living dead. These were the ascetics who had died to the world and who, like those in graveyards, represented an antithesis of the values of the human city. Indeed in later monastic practice, neophytes went through a quasi-burial ritual in which from prostrating on a funeral pall they rose, literally and metaphorically, with new names and a new identity.

Nevertheless there was some debate about the value of special places and of the growing habit of undertaking a pilgrimage to the tombs of saints as a spiritual discipline and means of personal transformation. In the late fourth century, St Gregory of Nyssa was already launching an attack. A 'change of place does not effect any drawing nearer to God'. St Gregory was particularly concerned to contrast Christian attitudes with pagan visits to temple shrines. In one sense he was defending the ubiquity of God's presence. In another sense he was defending the apophatic pole of Eastern theology – that God is not only beyond human language but is essentially inaccessible.[33]

Despite this ambivalent attitude, tombs, cults and pilgrimages played an important role throughout the Middle Ages in the

religious experience of the majority of Western Christians and
continue today. It was the virtue of the holy person, and divine
power acting through human agency, that mattered more than the
place itself. Bodies of saints were moved, stolen, and even broken
up in attempts to benefit from their virtue. If a holy person lived
and worked in a particular area, that created an indissoluble bond
and gave the place and its inhabitants certain rights even after the
saint's death. Although holy people frequently moved away from
their places of origin, at the same time there frequently developed
an intense relationship between saints and particular places. This
especially applied to the location where the holy person died.
Gregory of Tours provided an interesting example in his stories
about the death of a certain St Lupacinus.[34] There was fierce competition for the saint's relics between the town where he died and
people from another neighbourhood. The inhabitants of the first
town based their case on the straightforward notion that 'this man
belongs to our people. He drew water from our river, our land
transferred him to heaven.'[35] Apart from the notion that what the
town had given the saint ought to be rewarded (the basic principle
of any social relationship), the main point was that living and per-
forming miracles in a given locality had made the saint belong there
and to its inhabitants. It was simply assumed that this bond was
indissoluble and that, therefore, St Lupacinus would be unable to
part with these people after his death.

The place of a saint after death also seems to have involved some
kind of 'force field' within the bounds of which the holiness of a
saint was able to be active. Thus there was a medieval tale associated with the shrine of St Martin of Tours. A monk possessed by
a demon could only be delivered from his torment when he was
brought to Tours. Here the power of St Martin overwhelmed the
demon. However, when the monk returned to his former place he
became possessed again.[36]

In conclusion, it is interesting that the doctrine of the Communion of Saints is one of the least developed of theological themes
in the West. Sadly the Reformation conflicts did not help this
situation because the issue became sidelined in an almost exclusive
attention on both sides of the divide to the validity or invalidity of

devotion to discrete saints and particularly their power as intercessors or mediators.

Physical places: cathedrals and shrines

The question of tombs dovetails with other motives in the
Christian tradition of building churches. Christianity began as a
sect within Judaism. Even after Christianity parted company with
Judaism towards end of the first century CE, it remained for a long
time as no more than a network of small cells of initiates who did
not need permanent buildings. The *Ekklesia* was the gathering of
those people called out of the world long before it ever came to
mean a building. In this context, the crucial period was the fourth
century when official cultic approval was given to Christian communities. The achievement of a public face for the Church brought
with it the need for public architecture. This initially took on a
functional character for congregational worship. At first, churches
were really no more than meeting houses. With the legalization of
Christianity under the Emperor Constantine came the building of
basilicas, the creation of public liturgy and the growth of artistic
decoration. The sites of major churches were often associated with
the tombs of holy people. The primacy of the Roman Church was
built upon the tombs of the Apostles both literally and symbolically. It is also worth recalling that churches were built for rituals,
such as baptism and the Eucharist. These symbolized the incorporation of the believer into the death and resurrection of Jesus
Christ. Thus the sacredness of church buildings cannot be separated from their role in facilitating the union between the Christian
believer and Jesus Christ as saviour.

A second major feature of Christian theories about place is therefore associated with the church building as a physical container for
the living Body of Christ. A Christian church is not meant to be
either a theatre or a classroom – a place *essentially* concerned with
ritual or with teaching. Churches are traditionally oriented towards
the East – towards the rising sun that symbolizes the divine Son of
the resurrection. The place of the main altar at the eastern end was
not an end in itself. For the altar's only purpose was to lift the eyes

of the community to Christ who sat at the right hand of God and
thus to drive the believer onwards on the journey of following
Christ. In a striking contemporary essay on art, architecture and
liturgy, Albert Rouet, Bishop of Poitiers comments, 'Sacred space
is that of God's nomads. This itinerancy is an important characteristic of those who seek God, of those who are members of the
People of God.'[37]

Without ignoring the theological symbolism of church buildings
or denying their rare power to shape communities of contemplation, Rouet notes that, beginning with an empty tomb, Christ's
place is now his Body. 'The Body of Christ is the place where
charity becomes visible.' For this reason, church buildings make
theological sense ultimately in relation to the human community
and the quality of discipleship that they encapsulate and enable.
For this reason,

> Ecclesial space also denies itself in a way. The hope which Christ
> gave to the Church cannot be contained in any limited geo-
> graphical spot. The Good News drives us beyond. The holiness
> of the person of Christ is shared, exteriorised, and communi-
> cated . . . Christianity is a religion without spatial limits.[38]

It is now widely recognized that there was a diversity of aesthetics, and therefore of theological symbolism, during the Middle
Ages, not least during the great age of the development of Gothic
architecture. This has been emphasized by such important studies
as those of Umberto Eco.[39] Gothic space has been characterized as,
among other things, dematerialized and spiritualized. It thereby
expressed the limitless quality of an infinite God through the
soaring verticality of arches and vaults which were a deliberate
antithesis to human scale. The medieval fascination with the symbolism of numbers cannot be ignored either. The basic three-storey
elevation of Gothic form (main arcade, triforium and clerestory)
cannot be explained purely by progress in engineering. Both
Rupert of Deutz and Abbot Suger in the twelfth century drew
explicit attention to the Trinitarian symbolism of three-storey elevation. However, later Gothic buildings, such as King's College
Chapel in Cambridge, are notable for another typically Gothic

characteristic. The stone walls that support the chapel have been pared down to a minimum and replaced by vast expanses of glass. The stories in the windows could teach the worshipper much about the doctrine of God and of salvation but there was also a sense in which glass expressed what might be called a 'metaphysics of light'. God was increasingly proclaimed as the one who dwelt in inaccessible light yet whose salvific light illuminated the world.[40]

Until recently there has been an unbalanced concentration on a Dionysian 'metaphysics of light'. In fact it is now considered that Augustinian aesthetics played at least as important a part in the monastic theology of a person like Suger as the theology of Pseudo-Dionysius. And even the Dionysian elements are often affected by Augustine's thought.[41] Hence *harmonia* or a fitting order established by God is a central theme. This fitting order both refers to the building and to the worshipping community that it contains.[42] Abbot Suger, the great twelfth-century theorist of the birth of Gothic architecture at Saint-Denis in Paris, referred to 'perspicacious order' as the key to his vision for the building – and *ordo* is the characteristic word in Augustine for the harmonious beauty of the cosmos.[43] There may be various ways of understanding 'integration' when it comes to architectural style and buildings. An Augustinian approach would certainly begin with the fundamental understanding of the Church as a community of people, of the faithful who make up the Body of Christ. This is the *tabernaculum admirabile*, the 'wonderful tabernacle' of Augustine's sermon on Psalm 41 (in the Vulgate) within which one attains to God:

> 'I will go,' [David] says, 'into the place of the wonderful tabernacle, even unto the house of God!' For there are already many things that I admire in 'the tabernacle'. See how great wonders I admire in the tabernacle! For God's tabernacle on earth is the faithful.[44]

However, this *tabernaculum admirabile* has a *locus*, a physical place where it is both shown forth and continually reinforced. This place is, first, the liturgy, particularly the Eucharist, and then the building that contains this action. Thus the building in the mind of someone like Suger should evoke wonder, be adequate to its purpose of worship and point beyond itself to the ultimate 'house of God'. The building is a doorway or access point and its harmony is represented not simply by geometry or architectural coherence but by the degree to which it fulfils this function. That is its beauty, its *harmonia*.

This mixture of Augustinian and Dionysian elements seems to have been characteristic of Abbot Suger. This has led to a rereading of the theology Suger employed in reconstructing the Abbey of Saint-Denis, a building considered central to the development of all subsequent Gothic cathedrals. In fact Suger is replete with Augustine's sense of *harmonia*. For Suger, the inner meaning of an Augustinian theory of signs always pointed beyond to the 'more'. It was not so much a question of an aesthetic grounded in the physical beauty of the building as a higher sense of beauty that *necessitated* a transition from the material to the spiritual. Thus Suger actually quoted Augustine in the inscription on his great West doors.

The implication was that the higher understanding would 'see' the physical door of the building as the 'door of Paradise'. Suger summarized here both his theologies of light and of the great church as symbol of paradise:

Portarum quisquis attollere queris honorem
Aurum nec sumptus operis mirare laborem
Nobile claret opus sed opus quod nobile claret
Clarificet mentes ut eant per lumina vera

Ad verum lumen ubi Christus ianua vera
Quale sit intus in his determinat aurea porta
Mens hebes ad verum per materialia surgit
Et demersa prius hac visa luce resurgit.

Whoever you are, if you seek to extol the glory of these doors
Do not marvel at the gold and the expense but at the craftsman-
    ship of the work.
Bright is the noble work; but, being nobly bright, the work
Should brighten the minds so that they may travel through the
    true lights . . .
To the true light where Christ is the true door.

The gold door defines the manner in which it is inherent.
The dull mind rises to truth through what is material
And, in seeing this light, is resurrected from its former submersion.

There is a passage in Chapter 5 of Suger's *Libellus Alter De Consecratione Ecclesiae Sancti Dionysii* that explicitly links the 'admirable tabernacle' of the faithful in the Body of Christ to the building. The latter sets forth and makes concrete this family of believers and leads them on to Paradise:

> The whole building – whether spiritual or material – grows into one holy temple in the Lord. In whom we, too, are taught to be builded together for an habitation of God through the Holy Spirit by ourselves in a spiritual way, the more loftily and fitly we strive to build in a material way.[45]

Part of Suger's intellectual formation that lay behind his 'reading' of a church building was sacramental theology. What integrated and gave harmony to the building was Christian *practice*. As physical places, churches were locations of liturgical assembly. This is not merely a mechanical or practical issue – for example, associated with the development of passageways for processions or placing windows so as to light the critical liturgical focuses such as the high altar.[46] One of the key features of the theology of Suger and the cathedrals, and even the Dionysian theology propounded by Richard of St Victor, is that what is material is necessary to draw humanity upwards to the heavenly realms. To speak of cathedrals as a microcosm of the macrocosm, the cosmic order, is more than simply a question of representation but is a vision of Christian *practice* that acts as a medium of transition and transformation.

It is also important to add that medieval people had an integrated world-view rather than a differentiated one. They conceived of an ultimate unity in the universe. The conception of the Gothic cathedral is a good example. Every detail in the building recapitulates a grand architectonic design of the structure as a whole. This reflects in stone a conceptual approach to life in which the whole is somehow reflected in each part. This would apply to each separate chapter of a theological treatise such as the *Summa Theologiae* of Aquinas. Or it might involve a perception that every event in human history exemplifies something in sacred history. For the medieval person, a microcosm is simply a replica of the macrocosm. In antiquity people were not able to break out of the sphere of natural existence or stand up for themselves against the natural environment. There was a dependence on nature and an inability to see it as an object on which one could act 'from the outside'. This experience finds a cultural reflection in the idea of an inner analogy between humanity as microcosm and world as macrocosm – both having the same construction and elements.

Suger was clear that material realities were necessary rather than accidental to an apprehension of the divine. Every element of the building could be interpreted in reference to a higher meaning. 'Those who criticise us claim that the sacred function needs only a holy soul and a pure mind. We certainly agree that these are what principally matter, but we believe also that we should worship through the outward ornaments of sacred vessels . . . and this with all inner purity and with all outward splendour.'

Suger and the Abbey of Saint-Denis were undoubtedly influenced, even if not exclusively, by the works of Pseudo-Dionysius. For one thing, he was supposedly buried there and the monastery preserved the Greek text of his *Theologia Mystica*. An important element of Dionysian theology is the concept of light. God can be spoken of especially as light:

> Light comes from the Good, and light is an image of this archetypal Good. Thus the Good is also praised by the name 'Light', just as an archetype is revealed in its image. The goodness of the transcendent God reaches from the highest and most perfect forms of being to the very lowest. And yet it remains above and beyond them all, superior to the highest and yet stretching out to the lowliest. It gives light to everything capable of receiving it, it creates them, keeps them alive, preserves and perfects them.[47]

Everything created stems from that initial uncreated light. Each receives that light as appropriate to its place – a place according to the ordered hierarchy of beings established by God. The cosmos

was a kind of explosion of light and the divine light united everything, linking all things by love and with Love. There was, therefore, an overarching coherence. This outward movement of the
divine into the cosmos was complemented by a gradual ascent or
movement back towards the source of all things and all light.
Everything returned by means of the visible, from the created to
the uncreated.[48] The principle theme was the oneness of the universe. Having said this, we need to be careful. An overemphasis
on this theology of light is, it seems to me, to misunderstand
Dionysius. We may describe God as Light, yet, according to
Dionysius' own principles (especially as taught in the *Theologia
Mystica*), we must also deny that God *is* anything. God is *not* this,
not that – not even Supreme Light. In Augustinian terms, light is
the sign that may especially draw us to higher realities. But that
reality is ultimately beyond all conception:

> Trinity! Higher than any being,
> Any divinity, any goodness!
> Guide of Christians
> In the wisdom of heaven!
> Lead us up beyond unknowing and light,
> Up to the farthest, highest peak
> Of mystic scripture,
> Where the mysteries of God's Word
> Lie simple, absolute and unchangeable
> In the brilliant darkness of a hidden silence.
> Amid the deepest shadow
> They pour overwhelming light
> On what is most manifest.
> Amid the wholly unsensed and unseen
> They completely fill our sightless minds
> With treasures beyond all beauty.[49]

The theology of both Dionysius and Augustine was organized
around the Trinity. This was a symbol of God's creative outpouring into the cosmos and yet, at the same time, a symbol of the
impossibility of defining the ultimate nature of God. Gothic
portrayals of God, after the time of Suger, also focused on the

joining of God with human nature. As a result, one of the greatest symbols of the doctrine of the Incarnation and of the humanity of Jesus, the Virgin Mother, was situated at the heart of the icono-graphy of cathedral glass. Scenes of the Annunciation, the Visit-ation and the Nativity were found on the decoration of high altars. The Christianity embodied in Gothic was built on a theology of God as both almighty and unknowable yet incarnate and revealed. Gothic portrayed humanity as graced with divine illumination.

It has been commented of the medieval visitors to cathedrals that 'they were the enraptured witnesses of a new way of seeing'.[50] So what is this new way of seeing? Cathedrals in a sense contained all the information in the world and about the world for those who knew the codes. The way of seeing involved visual and other aids that drew the participant from self-awareness to God as the 'object' of a higher vision. Richard of St Victor in the twelfth century described the modes of vision in his commentary on the book of Revelation. There were four modes of vision divided between bodily and spiritual. In the first, we open our eyes to what is there – the colour, the shapes, the harmony – a simple seeing of matter. In the second, we view the outward appearances but also see their mystical significance. The movement is from immediate perception to a deeper knowledge. In the third mode, we move to the first part of spiritual seeing. Here there is a discovery of the truth of hidden things – such as the writer of Revelation himself experienced according to Richard. Finally, in the fourth mode, we reach the deepest level of spiritual seeing – the mystical. Here one has been drawn through the other modes to a pure and naked seeing of divine reality.

The 'real' lay beyond the apparent realm of the senses. Abbot Suger understood the architecture of churches as a harmonising of opposites – the act of divine creation reflected in the church:

> The awesome power of one sole and supreme Reason reconciles the disparity between all things of Heaven and Earth by due pro-portion: this same sweet concord, itself alone, unites what seem to oppose each other, because of their base origins and contrary natures, into a single exalted and well-tuned Harmony.[51]

In fact the Christian theology of physical 'sacred places' such as great churches or places of pilgrimage is also essentially associated with people, living or dead, as the *loci* of the sacred. Pilgrimages were to the shrines of saints, and the great church was simply a space within which the living story of God's dealings with the human community could be told through architecture, glass, stone and the liturgical assembly. If the architectural order of the great church was a microcosm of the cosmic order, that order consisted of a hierarchy of beings rather than simply an impersonal geometrical pattern.

In social terms, the development of the great cathedrals was most obviously an urban phenomenon. It also represented an eschatological shift.[52] In the early Middle Ages 'the sacred' was represented most radically by individual ascetics or monastic communities who were largely sited in wilderness areas. The dominant image of heaven was not surprisingly the recreation of Eden paradise. After about 1150 the first major urban renewal since the demise of the western Roman Empire had a major impact on social and theological perspectives even though the major increase between 1150 and 1250 still embraced only some 5 per cent of the European population. In terms of biblical theology, there was a gradual move from the book of Genesis to the book of Revelation, from a Garden restored to a New Jerusalem. At the heart of the new cities grew up the cathedrals and the new Gothic style.

In the urban cathedral, heaven was not only invoked symbolically but also, as it were, brought down from heaven in the spirit of the book of Revelation, Chapter 21. To enter the cathedral was to be transported into heaven on earth by the vastness of the space, by the progressive dematerialization of walls with a sea of glass and a flood of light and by the increasingly elaborate liturgies in which, sacramentally, the living Church was united with the whole court of heaven. Guillielmus Durandus in the thirteenth century in his *Rationale divinorum officiorum* suggested, in reference to solemn processions through the church and up into the sanctuary, 'When entering the church while we sing we arrive with great joy in our [heavenly] fatherland . . . The chanters or clerics in their white robes are the rejoicing angels'.[53] For Suger the church building had to be more impressive than other buildings of the city. The treasures should evoke the splendour of heaven, and the liturgical officiants, like the blessed in heaven, would dress in silks and gold.

The art of the cathedrals celebrated an incarnate God and attempted to portray, evoke and invoke a peaceable oneness between Creator and creation. This was a utopian space in which an idealized harmony, to be realized only in heaven, was anticipated in the here and now. But it *was* idealized. As Georges Duby, one of the most distinguished French medievalists of the last fifty years, reminds us, 'Yet it would be a mistake to assume that the thirteenth century wore the beaming face of the crowned Virgin or the smiling angels. The times were hard, tense, and very wild, and it is important that we recognise all that was tumultuous and rending about them.'[54] The social symbolism of cathedrals was also ambiguous.

Harmonia or order tended to be conservative in its results. The perfect, harmonious community of course reflected current social hierarchies and values. So, for example, it has been noted that representations of heaven idealized in the art of cathedrals tended to reproduce rather than subvert the separation of laity from clergy and the peasantry from the aristocracy and the monarchy. Thus, on the West front of Chartres above the great door,

> elongated figures of 'saints' thinned out of the world to reach a God above, and stout, stocky figures of this-worldly artisans and peasants supporting with the sweat of their brows that other 'leisure class' who have all the time and energy for liturgies and mystical contemplation, point to a conception of spirituality indelibly sculptured in the cathedrals of our collective unconscious.[55]

Medieval religious buildings today stand in a completely different time frame and are experienced by a completely different audience. In fact there is no way back to 'real' cathedrals or a 'real' medieval audience. Since the late nineteenth century, there has developed a kind of nostalgic yearning to overcome the sterile separation of mechanical details in studying cathedrals and to recover the integrated cathedral. Shortly after the Second World War this approach was most substantially represented by Otto von

Simson's attempt, as an art historian, to suggest a new integrated spirituality after his appalling experiences of National Socialism and the disintegration of Germany.[56] However valid this quest may be in its own right, what is problematic is how far these attempts to capture a holistic vision for humanity, embedded in cathedrals, are genuinely historical insights or an expression of a late-twentieth-century spiritual *angst*.

One refreshing aspect of more recent work on cathedrals is the way it seeks to integrate art history with medieval studies and with theology. The result is a move away from the idea that buildings are simply monuments of pure architecture. The spirituality implicit in cathedrals is *not* an abstract notion of sacred places. It is critical to their theological interpretation that cathedrals are places of social connection and of community definition. A building without per-formance is merely a form of abstract styling and, whatever else may be said, that never was the 'meaning' of Gothic space.

The Reformers' suspicion of place

At the end of this chapter, I want to mention briefly the ambivalence of the Reformation tradition towards place as a revelation of the sacred. The concept of place (above, below, between) has even been interpreted by some Protestant theologians as a distinguishing fea-ture of 'Catholic' sensibilities whereas 'time' (past, present, future) has been thought of as more characteristic of Protestantism. For example, the Swiss Reformed theologian Franz Leenhardt inter-prets Catholic and Protestant tendencies in terms of two strands of biblical tradition. The Abrahamic strand includes the prophets, St Paul and the Protestant Reformers. The primary symbol is the word, our critical sense is hearing and the fundamental spiritual dynamic is a movement towards an eschatological Kingdom. The physical world, images and ritual need to be questioned because they tend to divert people from placing their security in God alone. The Dutch pastoral theologian Hieje Faber has written in rather similar terms. He adds to Leenhardt's typology that the key values in Protestant sensibilities are prophecy and pilgrimage. In contrast, the Catholic tradition, associated with Moses or the Gospel of John, places its emphasis precisely on *being located*. The location is pri-marily in a human community which is the carrier of the tradition. The kerygma is mediated through place, local particularities and the sacramental space of community. Faber, in particular, empha-sizes that the Catholic tradition highlights God's essential *presence* in the here and there. God's glory is made tangible in particular contexts and holy places.[57]

Whatever the degree to which these generalisations make sense, classical Protestantism has sometimes been less happy than the Catholic tradition with the sacredness of places. This is not surpris-ing. It believed, first, in the unbridgeable gulf between the holiness of God and sinful creatures. It was suspicious, at least in theory, of any cult of saints with its emphasis on a theology of mediation and on the outstanding moral virtue of *individuals*. While a contempor-ary Protestant theologian such as Wolfhart Pannenberg admits that pre-Reformation spirituality included an emphasis on the immediacy of communion with God, he nevertheless contrasts a distinctive Reformation representation of immediacy with the pre-vious dominant system of mediation through Church, sacraments and devotion to the saints.[58]

Protestantism in the spirit of Martin Luther also concentrated on the *sacred community* of the Church and downplayed the physi-cal locations, such as church buildings, that might be understood as a *sacramentum* of that community. The sacred community was constituted by the Word being proclaimed and the sacraments celebrated in the assembly of believers. Remarks by the German Protestant theologian Rudolph Bultmann are revealing of this ambivalent attitude:

> Luther has taught us that there are no holy places in the world, that the world as a whole is indeed a profane place. This is true in spite of Luther's 'the earth everywhere is the Lord's' (*terra ubique Domini*) for this, too, can be believed only in spite of all the evidence.
>
> In the same way the whole of nature and history is profane. It is only in the light of the proclaimed word that what has happened or what is happening here and there assumes the character of

God's action for the believer ... Nevertheless the world is God's world and the sphere of God as acting. Therefore our relation to the world as believers is paradoxical.[59]

To be balanced, it does need to be said that there are other perspectives. Thus John Calvin seems to have been more comfortable at times than Luther with the notion that the world of natural and human places is a *theatrum gloriae Dei* – a theatre full of wonders in which God's glory becomes apparent. The *loci communes*, the ordinary places of the world itself, become the stage on which divine revelation is acted out:

> Meanwhile, being placed in this most beautiful theatre, let us not decline to take a pious delight in the clear and manifest works of God. For, as we have elsewhere observed, though not the chief, it is, in point of order, the first evidence of faith, to remember to which side soever we turn, that all which meets the eye is the work of God, and at the same time to meditate with pious care on the end which God had in view in creating it.[60]

These classical Protestant difficulties with some ways of talking about the sacredness of place remind us of a vital element in theological reflection. There is, if you like, the danger of 'cheap sacredness' that the contemporary theologian Rowan Williams has castigated as 'the rather bland appeal to the natural sacredness of things'.[61] Any workable theology of place must move beyond this naivety to contend with estrangement, with what is flawed and damaged in material existence. In other words, there needs to be an ethical emphasis in any reflection on the relation between place and the sacred. However, a purely ethical approach to place is not sufficient. There must also be a sacramental sensibility in which the particularities of places may point beyond themselves to the mystery of God.[62]

3

The Eucharist and Practising Catholic Place

While thinking about place in terms of theology, I have already hinted at one critical issue. In the light of the Christian doctrine of the Incarnation, spirituality is fundamentally concerned about how to live within a complex world of particularities. Our human 'placedness' is specified by God's commitment in Jesus Christ to the world of place and time. The particularity of the event of Jesus Christ 'permits' the placed nature, the always particular nature of discipleship. Yet, while we think of God as revealed at the heart of every particular reality, God nevertheless exceeds the bounds of the local and specific. The divine presence and action cannot be imprisoned in any contracted place or series of places.

Thus, in Christian theology there is an inevitable tension between the local and universal dimensions of place. Within every particular there is an impulse towards the universal, or what we might call 'catholicity'. Discipleship simultaneously demands a particular 'placement' *and* a continual movement beyond each place in search of an 'elsewhere', a 'further', an ever greater. A disciple of Jesus is one who is called to journey across boundaries and to exceed limits in search of the true catholicity of place.

I have called this chapter 'practising catholic place'. 'To practise' obviously emphasizes the way human actions construct or shape the world of places. 'To practise' also suggests a process of preparation for what is beyond the practice itself. Practice is orientated not only towards present action but also towards preparation for some kind of future completion. Christian discipleship may be described as practising in the here and now, and in hope and faith, the world that we believe is being brought into being.

There is a range of fundamental theological trajectories that may assist reflection on the nature of place and the sacredness of place. Sacramental theology is clearly pertinent, assuming by this something that is much wider than the technologies of Church sacraments. Equally, my remarks in Chapter 1 about the narratives of place, contested place and the unavoidable politics of place suggest that an ethical perspective is also critical. In fact I have opted to focus on what one might call the coinherence of sacramental and ethical perspectives. I am influenced by the renewal of Roman Catholic theology since Vatican II, especially, but not exclusively, in terms of the overall tone of such figures as Karl Rahner and Edward Schillebeeckx. My position is therefore that we cannot conceive of sacramentality as the 'eccentric' intrusion of grace, or godly space, into what is otherwise a profane world. We exist in an essentially sacramental universe or in graced nature. However, provoked by the work of some liberation theologians (on whom Rahner was an immense influence), my position is also that while sacramentality implies God's free self-disclosure and self-giving it also depends on a human response. Thus the question of the ethics of our sacramental practice is critical as well. An underlying question I wish to address is this. What really is catholic place and how is the Eucharist the practice of catholic place?

The catholicity of God

Catholicity is more than a purely ecclesiological concept. It is a much more fundamental religious perspective in that it offers us an understanding of the nature of God.[1] I would go further. The words 'catholic' and 'catholicity' can only be used in their fullest sense of God. While we must always be cautious about using spatial language in reference to God, we may think of God as the only truly catholic place.

What does catholicity signify if we begin our reflections with God? I would like to return briefly to the comments I made earlier about particularity. Any theology of place must engage unequivocally with particularity. For Duns Scotus whom I cited, all things in their very particularity participate directly in the life of the Creator. This makes of each thing a unique and irreplaceable expression of God's beauty as a whole. Each and every thing, person, and environment is called to 'be itself' or, better, to *do* itself with utter intensity and concentration and, thus, in terms of Scotus' approach to the doctrine of the Incarnation, to 'do Christ'.[2]

The doctrines of Trinity and Incarnation have implications for more than human identity. They imply a divine indwelling in all material reality and the revelation of God through the whole created order.[3] For the Christian, ultimate truth is not an abstract concept to be found in a dimension of existence that is 'no place' in particular. On the contrary, the Incarnation anchors human experience of the sacred firmly in the world of particulars. Ultimate truth must paradoxically be sought through contingent times and places. These have the capacity to speak sacramentally, beyond themselves, of God's presence and promise. What we sometimes refer to as the 'scandal of particularity', that God in Christ is incarnated within what is bounded and limited, is a guarantee that every particular time and every particular place is a point of access to the place of God.

In the doctrine of Incarnation, on which Scotus was so fundamentally dependent, God is to be discerned within each particularity. To the rather simplistic question, 'Where is God or what is God's place?' the Incarnation responds that God is no where in the sense of unbounded. However, at the same time, God may be said to act in the 'within' of all things. This answer excludes two rather important things. First, God is not to be conceived of as an outside will, an external force or power that operates alongside – indeed competes with – physical forces in the cosmos to push and pull and arrange events differently. Second, as Aquinas reminded us, we cannot speak of God being within things, people or situations as an aspect or dimension of their own inner constitution. However, we may speak of God in all things in the same way as we conceive an active principle to be present to what receives its action.[4] God's presence-as-action directly and intimately touches the within of each thing. God is the source of, and the goal of, each thing in its interior dynamism. In Scotist terms, God's life is so fruitful that it constantly, and inherently, expresses itself in the particularities of creation. In Christ, the outpouring of this divine life has as its end, or purpose, our deification or sharing in God's life. Rather than turn God's presence into a thing or object (rather like certain unhelpful ways of talking about 'grace'), we might want to say that God's presence is God's 'practice' of the superabundance of God's own self.

As the presence that acts in the within of all things, it is appropriate to think of God as 'making space'. I mean this in two ways. First, God may be said to ground everything in its own particular reality. This language of grounding is intimate and inclusive. It affirms that God is not bounded but encompasses everything that is. This is the easy side of affirming that God is the one who 'makes space'. However, second, God's active presence in the within of all things transgresses the boundaries of particularity. The universality or catholicity of God not merely grounds but expands the within of all things. This catholic space of God is subversive in that it not only makes space for the particular but also makes space at the heart of each particular reality for what is other and more than itself.

God is not in a protected place shrouded in a hermetically sealed ontological box. God is, rather, a shocking presence in a world of ambiguities. God both provokes an eschatological dynamism and, following the thought of Dietrich Bonhoeffer and the theologians of liberation, is a God who suffers on the cross at the heart of the world. This vision of God disturbs any tendency on the part of theology to settle for a comforting indwelling of God in the world of protected particularities. The catholicity of God's transcendence cannot be domesticated or controlled. This is the insight both of radical prophetic theology and of mystical theology.[5]

The presence of God acting in the within of all things is always strange and elusive, overflowing boundaries into what is 'other'. This excess, overflow and transgression confronts the human tendency to self-enclosure and to individual self-reference as the measure of everything and everyone else. There is a way of placing 'the other' on the margins in reference to the individual person as centre. God is the disruptive action within each person that decentres this illusory centre. David Tracy writes of God being the 'radical interruption' of versions of human history that oppress the marginalized. I would add that the catholicity of God is also a radical interruption at the heart of all individual lives that challenges self-containment.[6]

The famous dictum of Karl Rahner is that 'the "economic" Trinity is the "immanent" Trinity and the "immanent" Trinity is the "economic" Trinity'.[7] The inner life of God is to be understood not as essentially other than, but as graphically expressed in the whole economy of creation and salvation. There can be no effective separation between what God is and how God acts. God freely does what God is. God as Trinity is a space where the particularity of the divine persons is shaped by the interrelatedness of their communion.[8] Trinitarian language frees particularity from notions of radical separateness or discreteness.[9] Particularity in Trinitarian terms is radically open to what is other – indeed utterly dependent on what is other. The notion of God as distinct but not discrete persons-in-communion suggests a personal space for each that is, at the same time, and indistinguishably, a 'space for the other'.[10] It is this Trinitarian space that acts within all things affirming both the uniqueness of each particularity and yet, at the same time, its fundamental orientation to what is other than itself.

Catholicity and human place

Our reflection on catholicity moves from God as Trinity to the Church and world through the medium of Christology. The catholic 'space' that is shaped by God-as-Trinity finds expression in our time and space through the Incarnation. Thus, the focus of catholicity in space and time, without which it remains insubstantial and diffuse, is the person of Jesus Christ as the living fullness of God. In Christ, 'the whole fullness of diety dwells bodily, and you have come to fullness in him, who is the head of every ruler and authority' (Col. 2.9–10). And 'From his fullness we have all received, grace upon grace' (John 1.16).

In so far as the catholicity of God is mediated through Jesus Christ, it is important to note in reference to the New Testament that an important feature of Jesus' practice was to push people, not least those closest to him, away from familiar places into locations they found disturbing. To put it another way, the actions of Jesus redefined the nature of what was 'centre'. He regularly moved beyond the exclusiveness of the traditional Jewish land to reach Gentiles in outlying areas. So, for example, there is a suggestion of tough words being needed to force reluctant disciples into the boat to cross to Gentile Bethsaida in Mark 6.45. It is in places on the edge, and among those considered God-forsaken by many of his contemporaries, that Jesus knew his identity as Messiah must be revealed. He healed the demoniac on the East Coast of the Sea of Galilee in the land of the Gerasenes (Luke 8.26–39). He crossed into Tyre and Sidon to heal the daughter of a Syro-Phoenician woman and to commend her faith (Matt. 15.21–28). He healed in the Decapolis. In Mark 8.1–10 he fed a multitude on the *eastern or non-Jewish* side of the lake. 'Ever dragging his disciples away from the familiarity of home, he declares present the power of the kingdom in the alien landscapes of another land.'[11]

The community of believers that we call Church uses the word 'catholic' to describe itself in that it is the Body of Christ, the place of the Risen Jesus extended throughout human places and human history. Catholicity has the sense, among other things, of 'telling the whole truth' or expressing the complete story. That is, it has to do with the way in which the fullness of Jesus Christ's story, and the wholeness expressed *in* his story, is made real to us. In shorthand terms, catholicity is concerned with living in Christ or, to turn again to Duns Scotus, 'doing Christ'. However, catholicity is also always about expectancy, continuous transformation and about a process of becoming. Catholicity is not a characteristic that is ever finally possessed by any human group; it is always in process of becoming. In Michel de Certeau's terms, when we really *practise* place, this action refuses to be circumscribed but passes continually from one location to another.[12]

When we speak of catholicity we are neither referring simply to a literal geographic spatialization nor to some form of cultural homogeneity. Otherwise there would be many contenders for the title 'catholic' in political, economic or social systems. Even in a religious sense, a community is not catholic simply because it is represented in every town. Rather, the catholicity of a community is the demanding call to manifest God's own act of making space for all particular realities in their interdependency, we might even say in their *perichoresis*, and of transcending the boundaries of time and place as well as natural or cultural divisions between people.

So, catholicity does not mean an avoidance of the complex world of the particular. Its 'universality', on the contrary, is one not of abstraction but of the radically specific. The 'catholic' sense that God cannot be imprisoned in any contracted place implies that the divine must be sought throughout the *oikumene*, the whole inhabited world. If I may borrow once again from Michel de Certeau, the practice of catholicity drives us ever onwards to embrace the 'elsewhere', 'the other', the *semper maior* – the always more, the always beyond. The dynamism of catholicity pushes us forever to transgress boundaries and to exceed limits. Catholic place, in our contingent experience, can never be simply an arrival point but always implies a further departure. This is because catholicity does not mean simply what is ubiquitous but what is whole and complete. No human person or community can be that within the contingencies of time and place. Catholicity is what makes something full and what supplies totality.

When we speak of catholicity in our time and space, we are not addressing exclusively the structures of the institutional Church – although the word is frequently confined to an ecclesiological sense. Catholicity also expresses the fundamental value of achieving full humanity in the image of God. The word itself derives from the Greek *katholikos*, which means 'general' or 'universal'. However, that is not limited to 'worldwide' in a geographical sense. It is generally thought that the word *katholikos* has its roots in an adverb *kath' holou*. This means 'on the whole' or 'generally speaking' or 'in general'. This adverb again has connections with *kata holos*, meaning 'in respect of the whole', or 'that which is not partial'. At its roots, catholic is the opposite of sectarian – whether in a strictly religious sense or in its more general social or ethnic sense as 'narrowly confined to a limited and exclusive group of people'. Thus Augustine in his *Epistle* 49, n. 3, contrasts a catholic community with any group (he was thinking of the Donatists) that wishes to separate itself off from the general mass of people.

In recent writing there has been a number of interesting attempts to summarize the key characteristics of catholicity. An interesting way to begin is by means of a variation on the spatial theme. Thus catholicity has breadth. It is broadly inclusive, not bound to a single culture, and is opposed to divisiveness and individualism. Catholicity has length. It indicates a communion with every generation and cannot be limited to a particular historical period or narrative. Catholicity has depth. It permeates all dimensions of human nature and culture and is open to truth and values wherever they are to be found. Finally, catholicity has height. That is to say, its ultimate coherence is not achievable by human striving alone but is brought into being by our participation in the life of God.[13]

There is a number of fundamental catholic principles. These may be summarized under three headings: sacramentality, mediation and communion.[14] A sacramental sensibility understands the divine to be accessible through the human, the universal through the particular, the transcendent through the contingent, the spiritual through the material, the ultimate through the historical. We may want to affirm that the *fundamental* sacrament is Jesus Christ and, by extension, the community of the Body of Christ. However, a sacramental sensibility enables us to affirm that God's presence is active in the space of the *world*, not merely within a gathered and purified Church. This perception receives one of its classic expressions in Western spirituality in the concluding *Contemplatio ad amorem* of the *Spiritual Exercises* of Ignatius Loyola:

> I will consider how God labours and works for me in all the creatures on the face of the earth; that is, he acts in the manner of one who is labouring. For example, he is working in the heavens, elements, plants, fruits, cattle and all the rest – giving them their existence, conserving them, concurring with their vegetative and sensitive activities, and so forth.[15]

A sacramental imagination affirms that the divine eliminates the absolute dichotomy between nature and grace. This is not a matter of rose-tinted unreality – cheap sacredness – an affirmation, without qualification, that the world in the words of Thomas Traherne is 'a mirror of infinite beauty', 'the Temple of Majesty', 'the Paradise of God', 'the place of Angels and the Gate of Heaven'.[16] However, to refuse to acknowledge a deep dichotomy blocks any easy escape route from the dark side of space–time existence, taken by many groups from Donatism onwards, into a separatist utopia. There is no protected spiritual realm to hide behind. The human condition demands that people express their identity within materiality despite all its ambiguity. Because of this engagement rather than retreat, a sacramental sensibility forces Christians to confront the hard ethical questions posed by human selfishness as well as the radical orientation of human existence toward the divine. Thus, the sacramental sensibility that characterizes catholicity is inextricably forced to engage with disharmony, with the transformation of the world and with the call to reconciliation.

The second and third principles are mediation and communion. These are fundamentally corollaries of sacramentality. *Mediation* expresses a belief that God is available to us and acts for us in and through all created realities – our personhood, interpersonal relationships, society and the natural world. Once again, the human encounter with the divine does not take place in another protected space – human inwardness – whether an inward conscience or an inner consciousness. *Communion* emphasizes in particular that the encounter with God is not experienced in isolation but is made possible only through the medium of relationships whether interpersonal or social. This particular point raises important questions for how we understand the so-called mystical journey in Christian terms when mysticism is so often interpreted in terms of individualised interiority and isolated experience. To what extent is mysticism concerned with communitarian space – even political space? This is, in part, the focus of the next chapter. This does not imply that catholicity leads to a kind of totalitarianism that demands the suppression of individuality. Rather it concerns the coinherence of the personal and the collective. To put it more succinctly, the catholic nature of the Christian community is to be open to all humanity. The catholicity of Christian faith strives to penetrate all dimensions of existence. The universalism of catholicity involves bringing together many different aspects, even some that are acutely in tension, rather than uniformity.[17]

Although I have suggested that I would extend catholicity beyond its more common ecclesiological sense, it is nevertheless illuminating to compare two contrasting ecclesiologies with regard to Catholic identity. The first predominated within the Roman Catholic communion before Vatican II and the second took centre stage after the Council – not least in ecumenical conversations – although there is plenty of evidence of continued resistance to it. To be 'Catholic' in a pre-Vatican II world was to be universal and inclusive in a quite simple way. The institutional Church was a 'community fully in possession of itself'. This community was a *societas perfecta*. It was the place to which all mediation of the sacred was confined. It contained all that was needed for salvation. The failures of other institutions detracted in no way from its self-contained fullness. It was not in a process of becoming. Its membership was effectively co-extensive with the theological Body of Christ – although some honourable exceptions were hesitantly allowed. What was 'other' could safely be rejected and excluded without any sense of injustice or loss, for the 'outside' could neither add anything that was lacking nor offer any meaningful critique of the Church's integrity. At the heart of the institutional Church was a spiritual vision that transcended (or was intended to transcend) the vagaries and contingencies of local cultures. Post-conciliar concepts of catholicity, on the other hand, point to a rather different picture. The 'universality' of catholicity is expressed in terms of a single baptismal call to holiness and mission, yet one expressed in a variety of lifestyles, practices and ministries. The institutional Church is not to be thought of as a perfect society but as a contingent reality, albeit one with a sacramental identity, embedded in a glorious variety of human cultures. This necessarily makes the living reality of the Church in time and space partial at any given moment. The community might be said to have a vocation to catholicity in the grace of God. A vital aspect of true catholicity is therefore to be found, paradoxically, in a responsive attitude to what is other.[18]

The Eucharist as catholic place

Within our world of particular places, the catholicity of God is sacramentally expressed most explicitly in the *koinonia* of believers filled with the Spirit of the Risen Jesus and shaped by the practice of the Eucharist. The space of the Eucharist has a particular potency in terms not only of a tension between particular and universal place but also in terms of a process whereby the Christian family is called to bring into being a reconciled space at the heart of the world.

My preferred way of approaching the Eucharist as 'catholic place' is in terms of what I call 'ethical sacramentality'. This means that I reject a bland sacramentality that *merely* emphasises that humanity lives in the midst of a graced world. It is not that this assertion is untrue. It is simply that it seems to me insufficient. I suggest that real sacramentality is extremely demanding. It insists that Christians pay an ethical and political price for being a people who espouse a sacramental world. Sacramental theology and ethics are not distinct fields that have to be brought together. In reference to place, for example, it is sometimes implied that a sacramental approach and an ethical approach are alternative and, indeed, competing perspectives. To my mind they do not merely complement each other but actually coinhere. A truly sacramental view is necessarily an ethical view and vice versa.

One of the most significant American Roman Catholic moral theologians, Charles Curran, has commented in writings on catholicity that 'All must agree that the Church constitutes a community of moral conviction.'[19] By this he means that moral convictions necessarily flow from the fundamental nature of the Church as a community committed in faith to Jesus. I would go further than this to make explicit what I think Curran's approach actually implies. That is, that the very nature of the Christian community as the Body of Christ is to 'make space' in love for the other. In that sense, the Church *is, or is called to be, an ethical space* rather than, in a kind of detached way, to 'have moral convictions'. Its very life is an ethical practice, that is, a practice of making place for the fullness of all things and all people. To put it in other terms, in its Trinitarian life, the Christian community (not exclusively but as sacrament of human community) is a place orientated to a universality that does not depend on eliminating particularities.

The Eucharist as ethical space

The Eucharist is not simply a practice of piety but the enactment of the special identity of the Christian community. As such it is *ethical* practice although not simply in the superficial sense that it provides an opportunity for a didactic form of moral formation.[20] It seems to me that the link between ethics and the Eucharist is intrinsic rather than extrinsic.[21] Ethics is never merely a question of behaviours or practices. It embodies a way of being in the world that is appropriated and sustained fundamentally in worship, especially the Eucharist. Conversely, the eucharistic enactment of Christian identity necessarily opens the believing community to appropriate ways of living in the world of places and people. This may be expressed in the two related themes of reconciliation and solidarity.

As the ground-breaking 1982 document of the World Council of Churches on Baptism, Eucharist and Ministry made clear, a renewed eschatological emphasis in contemporary sacramental theology provides one of the most substantial groundings for the link between a sacramental and an ethical view of reality. In its eschatological dimension, the Eucharist shapes an ethical space at the heart of what is existentially a flawed and ambiguous world:

It is in the eucharist that the community of God's people is fully manifested . . .

The eucharist embraces all aspects of life . . . The eucharistic celebration demands reconciliation and sharing among all those regarded as brothers and sisters in the one family of God and is a constant challenge in the search for appropriate relationships in social, economic and political life . . . All kinds of injustice, racism, separation and lack of freedom are radically challenged when we share in the body and blood of Christ . . . As participants in the eucharist, therefore, we prove inconsistent if we are not actively participating in this ongoing restoration of the world's situation and the human condition. The eucharist shows us that our behaviour is inconsistent in face of the reconciling presence of God in human history: we are placed under continual judgment by the persistence of unjust relationships of all kinds in our society, the manifold divisions on account of human pride, material interest and power politics and, above all, the obstinacy of unjustifiable confessional oppositions within the body of Christ . . .

The eucharist opens up the vision of the divine rule, which has been promised as the final renewal of creation, and is a foretaste of it. Signs of this renewal are present in the world wherever the grace of God is manifest and human beings work for justice, love and peace. The eucharist is the feast at which the Church gives thanks to God for these signs and joyfully celebrates and anticipates the coming of the Kingdom in Christ.[22]

A sacramental sensibility is not affirmative of the sacredness of the world of places in some naïve way. It also has a transformative element. The world may be a gift of God but it is also a place at odds with itself. It is both glorious and profoundly disturbing but never neutral. The classic Protestant difficulty with the sacramentality of place is a significant reminder of something vitally important. A true sense of the sacredness of the place that is our world has to be carved out of a process that begins with estrangement, passes through surrender and finally reaches recreation. A sacramental perspective on reality demands more than a simple recognition of a sacredness that is 'there'. There must also be a reordering of the existential situation in which we are. To live sacramentally involves setting aside a damaged condition in favour of something that is offered to us by grace for 'where we habitually are is not, after all, a neutral place but a place of loss and need' from which we need to be relocated.[23] Part of this damaged reality consists of our flawed identities – whether these appear to enhance us as people of power or to diminish us as people of no worth. The transforming place of the Eucharist demands that the presumed identity of all people be radically reconstructed. This necessitates honest recognition, painful dispossession and fearless surrender as a precondition of reconciliation. Such reconciliation clearly does not come cheaply.

As I have suggested, the ethical 'space' of the Eucharist is the practice of godly space and therefore involves a quest for universality. In terms of Michel de Certeau's conception of 'spatial practices', the Eucharist is a practice of resistance to any attempt to homogenize human places. The Body of Christ, which the Eucharist shapes, is thus a place that critiques human totalitarianism. It rejects a detached universalism. In *this* place, people are not opposed or juxtaposed but precisely allowed space to be identified as who they are.[24] Their value is not, in Scotist terms, as merely particular instances of a general type. The Body of Christ is truly catholic to the degree that each member is able to 'do' or 'practise' themselves in all their specificity. The Body of Christ makes space for an infinite variety of stories as those who follow after the perpetually departing Jesus walk towards their unique destinations. Yet, at the same time, disciples are to walk the world–space in a way that also transcends the limitations of our bounded selves in order to reach out within time and space to the catholicity of God.[25]

To practise the Eucharist, to enter this space, implies a radical transformation of human 'location' such that it is no longer to be centred on the individual ego but discovered in being (if I may borrow a classic phrase from Ignatian spirituality) a-person-for-others. Costly reconciliation embraces an ethic of responsibility. Responsibility, in eucharistic terms, is not so much 'responsibility *for*' the other, which has echoes of one-sided authority or hierarchy. It is a 'responsibility *with*' others in a common life within the Body of Christ. This includes both a sense of co-responsibility and a mutual calling each of the other *to* the responsibility and obligations of love.

A place of reconciliation

So, at the heart of an ethical sacramentality, expressed in a theology of the Eucharist, is the critical theme of reconciliation. The Eucharist makes space in the world of times and places for God's reconciling presence in Christ to find a prophetic voice and to become effective in the community of the Body of Christ – the utopian carrier of the vision of the world's future. What is radically

Christian in affirmations about God is the notion of 'love'. This is not sentimentality nor is it a mere statement of simple relationality. For the measure of what is implied by the notion that 'God is Love' is the history of Jesus Christ, particularly as revealed in his kenotic self-giving. God's reality is, in Christian terms, most radically revealed in suffering for the other, in loving to the end, in remaining at the place of the cross. Purely human love can never be of this quality, yet the measure of the call by God for humans to love the other is the practice of the Eucharist that draws the believer into the mystery of Trinitarian life.[26]

Words are important. In the social and political world, reconciliation is often interchangeable with conciliation and accommodation. However, conciliation, much used in peace negotiations and industrial arbitration, is associated more with pacifying or placating. Accommodation enables pragmatic, negotiated arrangements based on compromise. Reconciliation is much more costly because it goes much further and much deeper. The *Oxford Dictionary* definition not only speaks of restoring harmony and concord but also, interestingly, suggests the reconsecration of desecrated places. All of the studies on human reconciliation I have read over the years from social, political, psychological and theological perspectives emphasize as *the* critical factor that it can only take place *between equals*. It is, therefore, the product of a process of *making equal space for 'the other'*.

Of course, reconciliation is a process over time rather than a single miraculous moment. With remembering comes recognition, not least the recognition by all of the destructive forces around and within them – whether that is unacknowledged guilt or destructive hate. Reconciliation also then demands repentance by all of those attitudes and actions that promote the exclusion or the diminishment or the demonizing of the other. Only when there is substantial repentance can there be effective forgiveness. There needs, too, to be a commitment to refusal. What I mean by refusal is that within the process of reconciliation all people need to learn how to refuse to participate in structures or behaviour that violate the other in whatever way. Because reconciliation is an ethical task, there is also a need for restitution after repentance. 'Restitution' need not be economic, but includes such things as restoring the value and identity of the other or enabling the empowerment of others. Because restitution involves the establishment of a just situation, this will always entail loss for some so that others may gain. Restitution is also bound up with reconstruction – the reconstruction of a quite different world of discourse and practice.

A place of reconciliation is not a melting pot but creates a space in the world of places for diverse voices to speak and to be heard. Most of all, a space of reconciliation invites all who enter to make space for the other, to move over religiously or socially, to make room for those who are unlike, and in that process to realize that everyone has become something different.

The Eucharist as reconciling space

The Eucharist is a space of reconciliation. To live eucharistically as a way of 'practising everyday life' involves an act of commitment. Every time the Eucharist is celebrated the participants commit themselves not to succumb to despair in the midst of the world's misery but to convert their time and place into a laboratory of ultimate hope. To celebrate the Eucharist also commits people, even more radically, to cross the boundaries of fear, of prejudice and of injustice in a prophetic embracing of other people, without exception, in whom we are challenged to discover the Real Presence of an incarnate God:

> Reconciled in the Eucharist, the members of the body of Christ are called to be servants of reconciliation among men and women and witnesses of the joy of resurrection. As Jesus went out to publicans and sinners and had table-fellowship with them during his earthly ministry, so Christians are called in the eucharist to be in solidarity with the outcast and to become signs of the love of Christ who lived and sacrificed himself for all and now gives himself in the eucharist.[27]

To live the Eucharist, as liberation theologians constantly remind us, necessarily involves existing in a state of tension. We celebrate fullness and feasting in the midst of a world in which the existential location of most Eucharists is one of acute deprivation.[28] There is a profound and painful paradox in proclaiming plenty in the midst of want. The Eucharist celebrates the destruction of the boundaries of mutual separation and exclusion. 'As many of you as were baptized into Christ have clothed yourselves with Christ. There is no longer Jew or Greek, there is no longer slave or free, there is no longer male and female; for all of you are one in Christ Jesus' (Gal. 3.27–28). Yet the location of most Eucharists is one of social, political and religious exclusions. A vital aspect of the eschatological dimension of eucharistic space is that it points to a conclusion, an ultimate meaning, that judges radically all human systems of exclusion whether these are modest or great.[29]

A most important element in place as a human construct is memory or, more precisely, multiple memories. Eucharistic place is very much a landscape of memory – not least of ambiguous and conflicting memories. Beyond the immediate participants, there are wider and deeper narrative currents in any eucharistic celebration. There is one central narrative, that of the events of God's revelation and presence in Jesus Christ that enables all narratives to have their space and at the same time reconfigures them. This makes space for those who have been given no place in public history to find a voice. It makes space for a new history that tells a different story beyond the selectivities of tribalism or sectarianism. This means that every other human story in the place is to be understood in terms of God's saving work. The difficult reality is that God's narrative, expressed in the Jesus events, highlights the ambiguities, injustices and pain in human narratives.

In fact, in Ricoeurian terms, the eucharistic narrative in a radical way makes a place for the risky business of creating human solidarity and for changing the status quo. It makes space for unheard narratives to be told and for the history of suffering and exclusion to demand redress. A place of reconciliation makes space for memories that refuse to remain silent. To forget is to exonerate or to excuse and that would be neither reconciliation nor forgiveness. Eucharist, sacramental-ethical place (despite our various human attempts to regulate and control it) engages a power beyond the ritual enactments themselves to become a space of alternatives that prises open an elitist history to offer an entry point for the oppressed, the marginalized, the excluded. The eucharistic action, according to its own inner logic, is the most public and also the most catholic space that there is in the contingent world of space and time. There is a perpetual and uncomfortable tension between this sacramental practice of reconciled place and all the many efforts of Christians to resist the logic of reconciliation.

At the heart of a theology of reconciled place must be the belief that human identities and places are determined by God rather than by social or economic networks or obligations. In the redefinition of personal and collective human identities brought about in baptism, Christian disciples are bound into solidarity with those they have not chosen or whose presence they have not negotiated and indeed would not choose of their own free will. Consequently, the new community, the new world, spoken of in eucharistic place is deeply disturbing of any humanly constructed social order.

The Eucharist does not simply bind individuals to God in a vertical relationship or bind people to each other in another kind of purely social construct. We are bound to one another *en Christo*. And Christ, who is the head of the body, is to be found persistently on the margins in those who are the least in the Kingdom of the world. The margins include those who are other, foreign, strange, dangerous, subversive – even socially, morally or religiously distasteful in our eyes. Yet the Eucharist insists that humans find solidarity where they least expect it and, indeed, least want to find it. We may recall the story of Francis of Assisi's encounter with the leper by means of which he passed from a romantic sense of God's revelation in the natural world to embrace the incarnate God in the excluded 'other'.

The Eucharist in its full sense, rather than purely as ritual action, encompasses two movements of surrender. God in Christ is freely surrendered into human hands. Equally the worshippers, by surrendering themselves to the inner dynamic of eucharistic practice, risk a displacement from their bounded social and spiritual locations in order, as members of Christ, to extend to what is alien and other the intimate hospitality of sharing food and drink. In its most fundamental reality, eucharistic practice is a participation in Jesus' own self-emptying. This practice involves committing oneself to a loss of free-standing autonomy and of the ability to dominate one's places, geographical or moral, and to a perilous opening up of hidden connections with what is strange and disturbing.[30] The Eucharist may be thought of as a liminal place of strange meetings that cannot be shaped by a phenomenalist geographics that maintains hard boundaries between inside and outside, centre and margins, universal and particular.

While the theology of the Church as the 'mystical Body of Christ' may be an improvement over institutional models,[31] it is right to point out some of its weaknesses. The most striking of these is its potential to act as a retreat into future place or spiritualized place away from engagement with the history and politics of human places.[32] On the one hand, the Eucharist is not an eschatological escape route to somewhere beyond times and places. The practice of ultimate reconciliation of all things in the Kingdom begins here and now in the subversive and purifying presence of Christ. It is, in the end, the social impact, even the political implications, of the concept of Real Presence that is far more critical than the almost exclusive historical preoccupation with its technology.

However, the Eucharist does not simply challenge and empower believers to reshape the social, economic and political world 'out there'. The eucharistic action is founded on the transfiguration, however we understand this, of the ordinary material of human feeding. We are what we eat. The practice of the Eucharist does involve the repetition of familiar words and rituals. However, the most challenging element is one of *recognition*. Who do we recognize as our co-heirs with Christ, and who are we increasingly able to respond to in the real presence of Jesus Christ? Recognition of what is close to us, yet excluded from us in ordinary place and time, is particularly challenging. We may recall the story of the Duke of Wellington, the victor of Waterloo and several times Prime Minister. The story goes that one Sunday as he left his pew to take communion in his parish church, a woman who was in front of him recognized who he was and stepped aside with a curtsey. The Duke is recorded as insisting that she continued ahead of him. 'Madam, at this Table at least we are equals.' The space of the Eucharist reshapes our social selves. But the 'at least' suggests to me that the good Duke would not have anticipated any realignment of social status once they had left the West Door. Eucharistic space so often does not overflow its initial location.

The core of the eucharistic doctrine of Real Presence, however one understands this, is the notion of *God's* critical recognition of us; God's affirming and life-giving gaze. All are incorporated solely because of God's recognition. The demands on those who practise the Eucharist may be, as a result, more elusive but they are also more powerful than any notion of solidarity based solely on a social theory, however inclusive or just it seeks to be.[33]

Real Presence also stands in judgment on all eucharistic performance. This implies more than the truism that the Christian Church as an institution is sinful and contingent. The statement addresses the fact that none of our eucharistic spaces is existentially a perfect politics. Real Presence confronts our exclusions and judgments. To put it more sharply, concrete eucharistic celebrations exist as ethical practice but are sometimes practised unethically. One might almost say that the twentieth-century movement of liturgical renewal, expressed for example in the sacramental theology of Vatican II, has at times been unhelpfully one-sided. The results have too often been ecclesiocentric. That is to say that liturgical discussion, and the reform of liturgical practice and texts, has brought about a rich re-engagement of the Eucharist with the theology of the Church. The problem, however, is a continued disengagement between a theology of the Eucharist as 'Church space' and the Eucharist as 'world space'.

However, the Eucharist has fundamentally Trinitarian roots as modern eucharistic agreements increasingly make clear. For example:

> It is the Father who is the primary origins and final fulfilment of the eucharistic event. The incarnate Son of God by and in whom it is accomplished is its living centre. The Holy Spirit is the immeasurable strength of love, which makes it possible and continues to make it effective.[34]

The Eucharist is to be set firmly within the total action of God in creation as well as redemption. In that sense, the whole world is included within the eucharistic celebration.

The result of this disengagement between Church space and world space is to diminish the practice of Christian identity that lies at the heart of eucharistic celebration. The encounter with the Risen Jesus may continue to be comfortably limited to the context of what the Bolivian theologian Victor Codina refers to as 'drawing-room communitarianism' whose aesthetic is not marred by any solidarity with the poor.[35] The commitment to solidarity, however, is inextricably bound up with the other key eucharistic themes of repentance and reconciliation. These imply not simply a transformation of sensitivities but a transformation of the practice of everyday living. It is perfectly possible, as Codina hints, to limit reconciliation and forgiveness to a magic circle – if not precisely of those who already like each other, at least to those whose social and cultural worlds overlap to a reasonable degree. Solidarity is a much more challenging concept. It pushes worshippers beyond the familiar boundaries of social, cultural and economic separations. Fundamentally, solidarity challenges every eucharistic gathering to find its relational identity beyond its own finite community. By following Jesus' command to 'do this in memory of me', to take bread and wine, to bless, break and share, the community does not merely put itself into the position of the needy recipient who accepts Jesus' gift of himself for the life of the world. Rather, the community, the Body of Christ, is also drawn into the same dynamic of being broken open for the life of all. The Christian community in its vocation of catholicity does not merely engage in a mission of pouring itself out for the world as a prolongation of the kenosis of Jesus Christ. Rather the Church *is*, most substantially, in its vocation of kenotic outpouring, in a transgression of its own boundaries, in overflowing excess, in leaving itself behind, in breaking itself open.

All human bodies receive their meaning within Christ's risen Body, which continually transcends the bounds of ethnicity, gender or class. The Christian tradition essentially presents Jesus Christ as the one who enters into an ethnically, socially conditioned and gendered place, the body, and then radically subverts these boundaries. In other words, the Body of Christ, of its nature, always exceeds and always transcends. The Body of Jesus Christ is, from the moment of the Incarnation, 'in the present' in the sense that it is ethnic, social and gendered in specific contexts of time and place. But it is also unstable. The instability is expressed in the Gospels in a number of ways such as the displacements of Transfiguration, Eucharist, Resurrection and Ascension. In the Transfiguration the Body of Jesus becomes transparent to divinity and a male Palestinian Jew attracts the disciples onwards to the God of all times and places in whom every desire is satisfied. In the Eucharist the risen Body of Jesus is transposed and extended beyond ethnicity, social class and gender into a food that is shared and that binds. The Resurrection reveals the ultimate displacement of the Body into what is beyond human grasp. Jesus in the Resurrection–Ascension dynamic becomes the complex, multifaceted Body of the Church.[36]

The risen Body of Christ extends, as it were, into the community of disciples who bear his name. This Body, therefore, as it gathers together eucharistically, commits itself to be the prolongation in the space–time world of the self-giving of God in Jesus Christ. A community that is truly eucharistic does not exist to be self-contained but, in the pattern of Jesus, to be taken, blessed, broken open and given for the life of the world. In other words, the Christian community comes into its true identity paradoxically not by being bounded but by becoming the space of Jesus Christ, a truly catholic space, committed to embrace the whole created order as part of the economy of God's salvation that makes loving space for the other.

This theology of solidarity does not undermine the ecclesial nature of the Eucharist. Henri de Lubac's aphorism 'the Eucharist makes the Church' remains valid. However, to be ecclesial is not the same as ecclesiocentrism. The eucharistic community is not merely in communion with its own inner life but also with the growing Kingdom of God, which is not co-extensive with the visible Church. The eucharistic epiclesis is not solely the transformation of the elements of bread and wine or of the worshipping community but extends outwards to a transformation of human history and the world of everyday places into the Body of the

Lord.[37] The Eucharist draws the community into the power of God that operates in history. In its prolongation of the work of Jesus Christ, the community 'tries to establish a living connection between the world of God and the human world'.[38] The 'corporeality' of the eucharistic community means that the action of Jesus Christ is embodied in the community and also that this community embodies Jesus Christ in and for history. To speak of the eucharistic Church as the Body of Christ also implies that, in the manner of Jesus Christ, the community is 'incorporated' into the material history of humanity. The eucharistic community is both a historical body and a mystical body. It tries, under grace, to make present in historical place and time something that is not palpable. What is 'more', what is complete, what is 'catholic' if you like, is present in the Church in what is seen. Yet it is present as something that surpasses the contingent, the particular, the local, the bounded.[39]

The most challenging, but rarely noted, element of a doctrine of Real Presence is the question of who and what Jesus brings with him into the eucharistic space. In receiving Jesus Christ the disciple receives at the same time all that makes up his Body. We find ourselves in communion not merely in some romantic way with the whole court of heaven, a communion of saints that somehow safely visits us from elsewhere and represents merely our pasts and our futures. We also find ourselves, if we dare to face it, in communion with the *totality* of the present as well. We know from the gospel narratives of the Last Supper (which we claim to re-enact whatever our tradition) that the catholicity of Jesus' act of incorporation included not only disciples like Peter who denied him but Judas who betrayed him. Those we prefer to exclude from communion with us in the world of public place are already uncomfortable ghosts at our eucharistic feastings. The story of the centurion Cornelius in Acts 10 raises an even more difficult question about nature of the Body of Christ. However shocking the fact that God had chosen someone who was Gentile rather than Jewish, the even more disturbing fact was that the Spirit fell on people who were not baptized. The doctrine of Real Presence is a kind of Trojan horse that outflanks our defences and brings into our space all that, from our limited perspective, we would rather exclude. That is undoubtedly particularly disturbing for those who have a well-developed sense of moral or ecclesiastical order. However, it is also ultimately encouraging, for it suggests that the boundaries of the Body of Christ, as practised by the Eucharist, do not depend on our human powers of incorporation.[40]

Finally, in the Eucharist there is a tension between local and universal. Every Eucharist, in which the Body of Christ is practised, is local and particular. Yet, at the same time it collapses all conventional boundaries not only of place but also of time. Every Eucharist is a transgression, a transit point, and a passageway between worlds. Participation in the Eucharist draws the community into a catholic narrative of place that embraces all contingent times and all specific locations.

The Eucharist as eschatological space

The reconciliation practised in eucharistic celebrations ultimately has an eschatological dimension. There are more and less helpful eschatologies. In the book of Revelation, it appears that a future world is created to replace a devastated old one. At the ultimate intervention, no place is found for the earth and sea. In fact, no place is found for place itself. So, in appropriating this perspective, the history of Christian apocalyptic often projects our true home into a redeemed future:

> But by bringing 'Jerusalem' down from a supernatural Above, it cut hope free of geography. Time, the dimension of transcendence in which biblical salvation history writes itself, lifts free in the Apocalypse of all former places. It is not, therefore, that the apocalyptic vision leaps to a state of immaterial placelessness; rather, it fantasizes a new space for new bodies. Yet the text conveys no tragic regret that something irreplaceable – the space embracing earth's lives, memories, and potentialities – would disappear. It does not grieve the passing of all known places.[41]

However, an eschatological horizon does not per se empty human events and places of significance by suggesting that meaning is to be found only in some indefinable future 'elsewhere' after the death of time. There are indeed over-future orientated eschatologies just as there are over-realized ones. To suggest blandly that God *has* redeemed the world despite visible evil, or to suggest that God's redemptive love merely awaits us on the other side of history, leaves us with a God cruelly detached from the here-and-now realities of human suffering and violence. The power of eschatology is to suggest that the beginnings of the Kingdom lie in history and the world of places, through a process of spiritual transformation. Yet this process is always to be completed and so there is perpetually an impulse to press on towards a final completion. The Kingdom of God is not to be conceived as a replacement world to annihilate the world of particular places but as this world fulfilled and completed.[42]

A balanced eschatology opens every 'present moment', and indeed history as a whole, to what is beyond it or more than the present instant. History, and the 'present moment', is thus not reduced in importance, let alone annihilated, but actually *expanded and enhanced*. Every moment not only contains the presence of the past but also the hope of the future. Every moment is also decisive. An eschatological perspective makes chronological time also *kairos* time. Each historical moment is an 'end moment' that makes whatever we do now an act of commitment to what is final or decisive.

One might say that every moment is 'eucharistic' in a broad sense. In each moment, Christians are exhorted to give thanks, to re-member the saving events of Jesus Christ, to anticipate in joyful hope humanity's ultimate destiny. Theologically and spiritually, celebrations of the Eucharist are moments of concentration and intersection that both gather all time, past, present and future, into the here and now and also bring all places, here and elsewhere, into the midst of each gathering of the community. Human time and human places are brought into transforming contact with the 'epoch-making event of Jesus Christ in its once and for all character'.[43]

Christians live proleptically, living in and through their practice the affirmation that the catholicity of God is bound eventually to be victorious and to prove to be all in all. It has been one of the most important shifts of thinking in the theology of the Eucharist, as a result of a remarkable degree of ecumenical agreement, that attention these days focuses less on the past in isolation and far more on anticipation of the future. Historical polemics about the Real Presence were essentially framed in terms of whether, and in what sense, the historical event of Jesus Christ and his saving acts were re-membered and made effective in our present time. This unbalanced approach to a memorial of the past diminished our understanding of the kind of place the Eucharist shapes. For the Eucharist also makes space for living 'as if'. It is a space in which God speaks to human beings and acts upon them out of their future. 'The Holy Spirit through the Eucharist gives a foretaste of the Kingdom of God: the Church receives the life of the new creation and the assurance of the Lord's return.'[44]

I described the Eucharist as a 'space' of *transitus* – where there is a passing over between worlds. Eucharistic symbolism is founded on death and rebirth. The breaking and sharing of bread is symbolic of a sharing in the power of Jesus' death that opens the doors to new life. However, practising the Eucharist also, by association, points the members of the community towards their entry *now* into a painful death – the death of a bounded way of being and a selective way of being social. To practise the Eucharist entails the risk of reshaping the place where we stand. We are covenanted now, through God, not only to fellow believers but also to a fellowship with the place of the world.

The overflowing abundance of God, present within each bounded eucharistic community, pushes it not only beyond its particular limits into the present *oikumene* in pursuit of its vocation of catholicity, but ever onward towards the ultimate catholicity of living in the fullness of God. The Eucharist shapes a community of memory in which the past, even the forgotten and unnamed, is not lost. It also shapes a community of hope that affirms that the past as well as what we are now, and do now, are and always will be an integral part of our future in the catholic space that is the mystery of God.

4

The Practice of Place: Monasteries and Utopias

Monasticism is essentially concerned with changing places, literally and metaphorically. At the heart of Christian theology, and therefore spirituality, is an invitation to enter a new world. Here and now, we are invited to become citizens of an imaginative world that reshapes who we understand ourselves to be and which is defined by the place of Jesus. In so far as the place of Jesus is now the Body of Christ, the *koinonia* of believers filled with the Spirit of Jesus, we may speak of the Christian community as the carrier of this imaginative world – indeed, as a proleptic expression of it. Nevertheless there is an unavoidable ambiguity. However much the Christian community proclaims this proleptic vision, it cannot avoid the fact that, like all human groups, it is flawed and conditioned. The Christian community is not so much the *exemplar* of Kingdom values as the *carrier* of them.[1] Historically, the monastic way has expressed the proleptic vision of Christian community in a particularly intensified form. Its specific purpose and power within the Christian community is to be a place that, while socially and culturally 'eccentric', is paradoxically where people seek to live out the imaginative world of the Kingdom in radical terms.

A movement to the margins

The 'desert' or wilderness in various forms has exercised a peculiar fascination throughout Christian history. This is especially true of the originators of the monastic movement in Syria, Palestine and Egypt in the fourth century:

Christian monasticism . . . originated in the kingdom of the scorpion and the hyena: a world of rock and heat. Several centuries later the biographers of holy men in north-west Europe also depicted their subjects as seekers after landscapes and environments which were correspondingly forbidding.[2]

One of the fundamental features of Christian monasticism is that it demands withdrawal. Why were physical deserts chosen for monastic communities? There have been many attempts to describe a special association between religious experience and 'the desert', whether literal or figurative. The theme of the 'desert' is common to many monastic texts. It is both a paradise, where people may live in harmony with wild animals, and at the same time a place of trial where ascetics encounter the inner and outer demons.[3] The 'desert' is frontier territory. Living on some kind of physical boundary symbolizes a state of liminality – of living between two worlds, the material and the spiritual.

The desert was originally associated with a theology of death and rebirth. It was to become the tomb before the tomb. It is recorded in Athanasius' life of Anthony of Egypt that he began his life in the desert by literally sleeping in a tomb among the bones of the dead. What more powerful symbol could there be of a loss of conventional human needs and values?[4] However, the underlying point is that through struggle, physical deprivation and submission both to a spiritual guide and to the realities of an empty landscape, the monk enters into a new life in Christ.

To move to the desert was both to journey towards a holy place and away from the place of sin, metaphorically speaking. To strive to perfect oneself morally involved a topographical displacement. So the attainment of a state of holiness was understood also in terms of a movement in space. The early ascetics in Syria, Palestine and Egypt, from the late third century CE onwards, deliberately sought out the empty spaces of the wilderness as the context for inner spiritual combat. One element seems to have been a desire to be freed from an identity provided by normal social ties. Monastic disengagement from the start was a social and political statement as well as a theological one. We cannot overlook the vital importance in the fourth century of 'social meaning'. The presence of heavenly power on earth expressed in the monastic life was closely related to an ascetic stance to 'this world' – represented by disentangling oneself from conventional social and economic obligations in favour of a reshaping of human relations.[5]

Egyptian ascetics did not live as close to the ordinary world as Syrian ascetics normally did. These differences were partly geographical. The traditional agrarian culture of the Nile valued a regulated life. Combined with the harsh realities of the Egyptian desert, this necessitated a spirit of co-operation for survival. Thus we see the gradual development of monastic villages. Syrian wilderness was never deep desert in the same way and was not so starkly separated from human habitation. Consequently the ascetics remained visible challenges near to where people lived. In their geographical isolation, the Egyptians tended to recreate the format of village community and became the *oikumene* (settlement) in the *eremos* (desert). The Syrians, while closer to everyday life, adopted more eccentric lifestyles.[6]

Western monastic rules differed in the strength of their regulations about separation. Almost universally, monastic property was situated so as to avoid contact with the outside world in pursuit of solitude. Even the moderate Rule of St Benedict recommended:

> The monastery should, if possible, be so constructed that within it all necessities, such as water, mill and garden are contained and the various crafts are practised. Then there will be no need for the monks to roam outside, because this is not at all good for their souls. (RSB 66:8)

In Chapter 4 of the Rule, 'The Tools for Good Works', withdrawal from the world is expressed in terms of practising different behaviour:

> Your way of acting should be different from the world's way; the love of Christ must come before all else. You are not to act in anger or nurse a grudge. Rid your heart of all deceit. Never give a hollow greeting of peace or turn away when someone needs

your love. Bind yourself to no oath lest it prove false, but speak
the truth with heart and tongue. (RSB 4:20–28)

Also, in Chapter 4:8, the scriptural injunction to honour father and
mother is changed to 'honour everyone', presumably because a
critical element of monasticism was the withdrawal from conventional social and family ties. However, the Rule is not rigid
about physical withdrawal as it lists among the work of the monks
items outside the enclosure such as relieving the poor, clothing
the naked, visiting the sick, burying the dead, going to help the
troubled and consoling the suffering (RSB 4:10–19). In terms of
'contact with the world', the other highly influential Western Rule,
the so-called Rule of St Augustine, is actually very moderate.
Thus, in reference to encounters with women, 'You are not
forbidden to see women when you are out of the house. It is wrong,
however, to desire women or to wish them to desire you.'[7]

Yet monasticism is essentially liminal to everyday human places.
The question is whether this liminality is intended to underline a
kind of utopian vision for redemption of the human city or whether
it condemns the city as irredeemable and offers itself as an alternative. The imagery of early monasticism often portrays a return to
a pre-Fall paradise state that might suggest that the human city is
the living symbol of fallen humanity.[8] There has always been a
strand in Christianity that understands humanly constructed
places as essentially expressions of humanity's sinful and alienated
nature. This has usually been based on the biblical story of Cain as
the founder of the first city. The first murderer becomes the first
city builder. Not an auspicious basis for the spiritual nature of built
environments. The city has been portrayed as at best a place of
ambiguity and at worst a representation of humanity's rebellious
heart and its rejection of God.[9]

The costly discipleship of monastic life implied making oneself a
stranger and outsider. Yet there was a paradox. The lives of the
great monastic founder, Anthony the Great, or of Simeon Stylites
the eccentric ascetic on his roadside pillar near Antioch, reveal that
holy men and women did not leave social or public roles behind
entirely. In fact, by standing (geographically and socially) outside

normal boundaries, the ascetic was accepted as a spiritual guide and social arbitrator.[10] It appears that monasticism, in its origins, should be viewed as having a prophetic role vis-à-vis the human city rather than simply as providing an escape route into an alternative, purified, universe.

Utopias

The concept of utopia expresses a world of imagination and desire. If the spiritual vision of monasticism may also be described in terms of such a world, an explicit anticipation of the Kingdom, some reflection on the nature of utopias may be illuminating. 'Utopia' is a made-up word based on Greek, courtesy of Sir Thomas More. It means 'no place' or 'nowhere'.[11] In popular understanding, 'utopia' often stands for 'an ideal place' but this is based on a misapprehension. Strictly speaking all imaginary places, good or bad, are utopias. To qualify as utopias, imaginary places need to be expressions of human desire – positive or negative. In other words, they build on a universal human longing. It is this quality that allows utopias to capture our attention and sympathy whether they offer practical possibilities or not. 'Utopia is where we store our hopes of happiness.'[12] To point the way towards a different kind of place we need to explore the dialectical relationship between the possible and the impossible – which the concept of utopia actually does.[13] Utopias are concerned with symbols rather than with actual potential places. Utopias are important because they allow us to explore the 'places of what has no place, or no longer has a place – the absolute, the divine or the possible'.[14]

To make way for the birth of a new world, utopias necessarily project an ending to the present world. What needs to be changed as we move from the present to the new is an endlessly controversial matter because all utopias carry the weight of our projections, shaped in particular times and places, of what we long for or loathe. The first problem in the creation of a new world is obviously people. Different versions of utopias somehow eliminate all the obnoxious or anti-social people. The danger of such visions, of course, is that they tend to undermine the varied tapestry of real human life. This implies that various spiritual communities that exist as real places, yet seek to anticipate the ultimate transformation of human existence, are open to a number of temptations. First they may wish to present to the world an impossibly idealized image of themselves. Second, they necessarily have a highly conditioned understanding of future transformation. Third, and as a consequence, they may induce a great deal of guilt in members who, as people in real time and space, fail to match this ideal. Finally, the combination of an idealized vision and an all too human membership sometimes leads to an unhealthy suppression of dissent and the delicious variety of human nature.

Religious utopias particularly speak of a world transformed where we may live in perfect harmony, free from suffering, divisions and injustice. In Christian terms, the ultimate positive utopia is paradise or the Kingdom of God and its opposite is hell.[15] Theologically speaking, eschatologies are therefore variations on a utopian theme. Needless to say, classical images of heaven, paradise and the afterlife enable the elimination of non-utopian people automatically!

A certain kind of utopianism can be an attempt to retreat into a blissful realm of separatist imaginings, pristine and free from human places compromised by sin and injustice. Such utopianism in fact denies our spatial location by pretending that we can retreat into a detached standpoint beyond immediate historical contexts. However, the truth is that this world, however compromised it is, is the only place we have. Alternative ways forward can only be constructed in the place where we find ourselves. Carol Lake's collection of short stories *Rosehill: Portraits from a Midlands City* set in the inner city of Derby, England, won the prestigious *Guardian* newspaper fiction prize in 1989.[16] One story, 'The Day of Judgement', explores what would happen if the people of Rosehill had to face the Last Day. God arrives in a new Noah's Ark but, to everyone's surprise, leaves the Ark in order to drink with the locals in a popular bar while the Ark sails away without him. Lake's point is that the Ark travels onwards towards a destination that represents a utopia based on what she calls the 'heartlessness of perfection':

The Ark is on the edge of the horizon now, its destination the heartlessness of perfection. Most of the inmates already know what they are going to find – endless fruit, endless harmony, endless entropy, endless endless compassion, black and white in endless inane tableaux of equality. It sails off to a perfect world; the sky has turned into rich primary colours and in the distance the Ark bobs about on a bright blue sea. (p. 119)

It is the streets of Rosehill that become intensely beautiful. Although this place represents estrangement, it also represents for those who stay (including God) the possibility of transfiguration and the Kingdom of Heaven built in the streets of Rosehill among those who are underprivileged.

Sir Thomas More in the early sixteenth century created the classic version of utopia. The status of the book is a mystery. Is it meant to be the portrayal of a genuinely ideal society, indeed a truly Christian society without the benefit of Christian revelation? Is it merely a witty joke? Is it something in between – a nowhere place based on human reason? We Christians who have the benefit of revelation manage to do much less well than the Utopians who can only rely on human reason. More's work is essentially rhetorical and to take it literally is to miss its subtlety. There is much in More's Utopia to admire but as much that falls short. It seems clear that Christian revelation is meant to provide the yardstick for what is to be taken seriously in More's vision. This would support details such as no poverty, no exploitation, no luxury and no idle rich in a land of happy, healthy and public-spirited people. However, revelation would (in More's likely understanding of it) reject such possibilities described in the book as complete religious toleration and euthanasia. Interestingly, More had strongly monastic sympathies and it is not fanciful to see this reflected in his description of elements of Utopian social life.

It is accepted that More is, in part, wrestling with the question of what it is to be a Christian. However, the question is posed within the parameters of Christian humanist values. More's two humanist friends, Erasmus and Guillaume Budé, the most perceptive early commentator on the book, both believed that it contained 'divine principles'.[17] Budé called Utopia *Hagnopolis* or 'holy community'. He describes the place as somewhere where miraculously people have achieved Christianity without revelation. Utopians possess in highest measure the virtues More described as Christian a decade earlier: simplicity, temperance and frugality. They also have faith in a God whose goodness and mercy they trust. Along with faith goes *spes* or hope of eternity. Also *charitas* or love. They were joined together by 'mutual love and charity', *mutuus amor, charitasque* (p. 224 line 8). Utopian institutions are described as the most prudent and holy (p. 102, lines 27–28). Nowhere does More make the connection between Christianity and Utopians more explicit than at the point where the European traveller, Hythlodaeus, actually converts some Utopians to Christianity. The Utopians are drawn to Christianity by Christ's teachings and character (p. 216). They do not become Christians because Christianity contrasts with their world-view but because it expresses their most cherished beliefs. That is, 'when they heard about Christ for the first time, they experienced not so much revelation as recognition'.[18] It is the common way of life of Christian discipleship that attracts them.

The rhetorical style familiar to Thomas More originates in the tradition of Renaissance humanism that influenced him and his intellectual circle. However, such a style also represented a return to the Bible (especially St Paul) and to the theology of the early Christian era (especially St Augustine). Rhetoric sought to communicate much more than information or argument. Its purpose was to evoke love, feelings and imagination and thus to move the human heart. *Utopia* is meant not simply to *instruct* but to *move* the reader to some kind of change of perspective and values. Rhetoric implies a deliberately contrived device to communicate to the reader something of importance beyond the device itself (in this case an imaginary narrative).[19]

Monasteries may be thought of as examples of utopias – nowhere places that express human aspirations and desires. Although utopias in a strictly formal sense are nowhere, places of the imagination, the idea merges at its edges with forms that may exist in real time and space. It seems to me that monastic community is not a concrete substitution for the human city, everyday life and ordinary

Christian existence. It is better to see monastic life as a rhetorical statement and as an act of resistance against a diminishment of imagination regarding our human future.

A key to this monastic way of resistance may be expressed in terms of 'living *as if*'. Monastic spirituality invites us to live as if the inner harmony, the interpersonal reconciliation, the social conversion of the Kingdom of God were actually the case. It is anticipatory but in the sense that practising the 'as if' is, in God's providence, an irreducible part of the Kingdom actually coming to be. So, it is recorded in the desert tradition that,

> A brother asked an old man, 'What shall I do, father, for I am not acting at all like a monk, but I eat, drink, and sleep carelessly; and I have evil thoughts and I am in great trouble, passing from one work to another and from one thought to another?' The old man said, 'Sit in your cell and do the little you can untroubled. For I think the little you can do now is of equal value to the great deeds which Abba Anthony accomplished on the mountain, and I believe that by remaining sitting in your cell for the name of God, and guarding your conscience, you also will find the place where Abba Anthony is.'[20]

The symbol of mountain and desert plays a central part in all religious systems.[21] They take on some of the qualities of utopias. The allure of these inaccessible and imaginary places is that they are archetypal expressions of longings and aspirations. Essentially, as in the last quotation, utopian places symbolize states of growth that we have not yet achieved but which we desire. Places or landscapes of the imagination have a unique capacity to describe the deepest longings of the human spirit. They are not so much untruths as exercises in faith. To put it another way, we tell these tales (or, as in the case of the 'text' of monastic life, we live the story) in order to give form to a world that is not yet here.[22]

Arguably the most explicit example of monasticism as utopia is the apocalyptic writing of the twelfth-century Italian Abbot Joachim of Fiore who, by his death in 1202, had become one of the most noted figures of his day.[23] Joachim believed in the imminent coming of a 'third age' of the Holy Spirit following on the ages of the Father and of the Son. This would be characterized by a utopia of monastic perfection. This is not so much monasticism as microcosm of the Church as the monasticization of the whole Church. History was a story of the gradual triumph of spirit over flesh and contemplation over mundane thought. There were to be two stages to this perfect realization of monasticism. First, during an imminent crisis, two new religious orders would arise to confront the Antichrist and his forces – an order of preachers and an order of hermits. Finally, there would be a future perfect monastic state of the Church. This is found in *The Book of Figures*. In Joachim's Twelfth Table, 'The Arrangement of the New People of God', the Church is divided into seven 'Oratories'. The seventh is the Oratory of St Abraham the Patriarch and of All the Holy Patriarchs. 'Under the name of this oratory will be gathered the married with their sons and daughters living a common life.' While living in separate dwellings they will share food and clothing and be obedient to the Spiritual Father (who governs the whole community). The biblical allusions make it clear that the model is the early Christians of Jerusalem of the book of Acts.[24]

Although Joachim is an extreme example of monasticizing the whole people of God, the classic monastic vision of an alternative world in some sense does involve everyone who visited or inhabited the enclosed space rather than merely the 'professional' ascetics. The Irish Columbanian tradition, for example, believed that all people were called from birth (not baptism) to the experience of contemplation. So, 'monastic' enclosures were places of spiritual experience, of non–violence, of education, wisdom and art. Within the enclosures there took place, ideally speaking, an integration of all elements of human life, as well as of all classes of human society.

The origins of Christian monasticism, if not always its concrete expressions, bear an interesting resemblance to 'heterotopias' in the writings of the French postmodern theorist Michel Foucault. Before developing this point further, it is important to note that Foucault himself did not understand the history of monasticism in this way. For Foucualt, monasticism was an example, comparable to prisons, of the idealization of 'order' and control. This was part of his post–Catholic characterization of Christianity as an institutional structure that manipulated and controlled by a process of silenc-ing.[25] Nevertheless, I believe it is possible to enter into an interest-ing conversation with aspects of Foucault's thought. His concept of 'heterotopias' is a case in point.[26] These are places that are 'outside of all places, even though it may be possible to indicate their location in reality'.[27] They exist in all societies as real places whose functions are 'other' than the norm. A 'heterotopia' is a heterogeneous rather than homogeneous site. In other words, in a single place, it juxta-poses several 'spaces' that are in themselves normally incompatible. In that sense, a 'heterotopia' is an 'impossible space'. It has the ability to transgress, undermine and question the alleged coherence or totality of self-contained systems. Utopias may thus be one form of 'heterotopia' although some utopian visions (for example, fascism) bear a strong resemblance to the structures of manipulation and control through silencing that Foucault was so concerned to subvert. 'Heterotopias' are 'something like counter-sites, a kind of effectively enacted utopia in which the real sites, all the other real sites that can be found within the culture, are simultaneously repre-sented, contested and inverted'.[28]

The possibility of a fruitful conversation between Foucault's 'heterotopias' and monasticism soon becomes apparent. Mon-asticism, too, is a counter-site, in some sense representing an ideal-ization of critical elements of human community yet, at the same time, effectively contesting the contingent reality of existing social structures. This conversation is taken further as we examine the key features of heterotopian place. First, there are two broad cate-gories. One consists of privileged or sacred places for people to undertake some kind of personal transition. Another consists of places of 'deviation' such as prisons or psychiatric hospitals. The interplay of the first category with monastic places is immediately intelligible. The second may appear questionable until we reflect on de Certeau's contention that there are connections between spiritual vision and 'madness'. Madness, of course, being a cultural construct based on particular understandings of normality and socialization. A 'heterotopia' can change meaning and even physi-cal location over time, depending on the culture within which it occurs. The monastic way, while always liminal, has expressed this liminality in a multitude of concrete forms. It has existed in the wilderness, beyond the boundary of the civilized and it has also existed provocatively in the midst of the city.

'Heterotopias' have a capacity to represent several different sites in one real place. Foucault offers the example of the oriental garden, a tiny place that nevertheless represents the whole world. Symbolic microcosms permeate Christianity as well. As we have seen, medieval cathedrals were deliberately constructed in this way. Monastic spaces have also been interpreted in this way, both in terms of their architecture and of the ordered form of life practised within them. This practice of life is deliberately shaped to reflect the Pauline transgression of boundaries that separate whatever is thought of as religiously or socially incompatible. 'In Christ Jesus you are all children of God through faith. For as many of you as were baptized into Christ have clothed yourselves with Christ. There is no longer Jew or Greek, there is no longer slave or free, there is no longer male and female; for all of you are one in Christ Jesus' (Gal. 3.26–28).

Foucault also understands 'heterotopias' as specialized spaces in real time that accumulate all time while themselves appearing to be outside time and its ravages. His own examples are museums and libraries, but it is reasonable to add sacred sites including monasteries. 'Heterotopias' tend to maintain a tension between isolation and accessibility with entry carefully controlled. A similar tendency in classic monastic sites – of withdrawal yet accessibility – has often been noted.[29] Finally, 'heterotopias' have a function in relation to all other places. Foucault suggests that they do this in one of two ways. Either they create an imaginative space whose purpose is to expose the illusory quality of the places we take to be normal, or they create a space that is ordered in a way that exposes our normal places as disordered. The parallel with the *theology* of monasticism is clear, even if particular historical manifestations of monasticism have not always had this quality.

The work of Michel de Certeau offers a different but equally interesting way of approaching the relations between utopian place and monasticism. For all his deliberate ambiguity as a prophet of transgression and rupture, it is clear that de Certeau continued to inhabit a place on the frontiers of Christian discourse.[30] His use of 'utopian' is not always entirely positive – for example in the context of an imposition 'from above' of a conceptual meaning for the human city that contrasts with the practice of urban living.[31] However, elsewhere 'utopian' becomes associated with a 'non-site', an 'emancipated space' for prophetic Christian discourse, a form of 'madness' where Christianity marks out its necessary difference on the margins of a secular organization of social practice.[32] Utopian thought and monasticism similarly exist on boundaries and are concerned with his distinctive themes of otherness and resistance. An effective utopian rhetoric never places 'otherness' completely outside the everyday, normal speech or space-time. Yet, as de Certeau proclaimed in many different ways, vision is always a form of madness. Utopias, to be effective, must also be marginal to the present particular place, to *this and that*. Indeed, more than marginal; they are essentially resistant to bounded particularities. Monasticism reminds us that authentic Christian discourse is always breaking apart, is always paradoxical. It can never be captured in institutional structures. Indeed, the origins of monasticism lie, at least in part, in a radical protest against the transformation of a post-Constantinian Church into a public (in the sense of socially powerful) body. Such a discourse 'speaks prophetically of a Presence who is both immediately felt and yet still to come, who cannot be refused without a betrayal of all language, and yet who cannot be immediately grasped and held in terms of any particular language'. In other, no less paradoxical terms, one is called to be faithful in our particularity to what is Other, Beyond, Ultimately Excessive.[33] Monasticism creates a space for this paradox, or 'rupture', to exist in relationship to everyday action.

De Certeau sought to speak in and for a world in which the Christian community is no longer the place of meaning. Its status as the exemplary form of human community has been increasingly overtaken by the secular state. How can we continue to believe in the absence of a distinctively Christian place? In the end de Certeau, after examining various models, seems to suggest that there is no *theoretical* construct available to describe Christian identity definitively. What is left is the age-old tension between discipleship (following) and conversion (change). The believer is one called to follow faithfully and to change:

> As the ecclesial 'body of sense' loses its effectivity, it is for Christians themselves to assure the articulation of this 'model' with actual situations. This 'model' refers to the New Testament combination of 'following [Jesus]' and 'conversion' . . . The first term indicates a going beyond which the name of Jesus opens up, the other a corresponding transformation of consciousness and of conduct.[34]

This tension is deliberately expressed in the Western monastic vows of stability, obedience and *conversatio morum* or journeying on a way of perpetual conversion. This tension cannot be resolved, but only embodied in practice where every movement forward is new. In response to the question, 'How would you define Benedictine spirituality?' an American monk once said, 'We ring the bell, we recite the prayers, we live the life.' How very de Certeau! In monastic terms, therefore, the narrative is not so much the written Rule as the variety of communities that live the Rule as a medium for maintaining this tension. De Certeau's project for the future of gospel living is a form of utopian utterance. The place from which Christ now speaks is an empty tomb.

Because such a lived human narrative can never reach a point of definitive arrival, it may appear to some people to have little or no power. For de Certeau, however, it is paradoxically the ability of the lived narrative (e.g. of monasticism) to question the *excessive* stability of places and systems of meaning that gives it its power. For de Certeau, the most powerful symbol of Christian meaning is the empty tomb. This is a 'no place' that disrupts truth, power, authority and any temptation to a definitive settling down *here* with the possession of the truth *of this or that*. The Christian call is to wander, to journey with no security apart from a story of Christ that is to be enacted rather than objectively stated and which is profoundly disruptive. The role of monasticism within Christianity and wider society has always been to stand as a disruptive act of resistance at the heart of our systems, our objective statements. But if the monastic tradition is to be truly evangelical, it must avoid trying to see itself as another way of being Church as a definitive place:

> The temptation of the 'spiritual' is to constitute the act of differ-
> ence as a site, to transform the conversion into an establishment,
> to replace the 'poem' [of Christ] which states the hyperbole
> with the strength to make history or to be the truth which takes
> history's place, or, lastly, as in evangelical transfiguration (a
> metaphoric movement), to take the 'vision' as a 'tent' and the
> word as a new land. In its countless writings along many different
> trajectories, Christian spirituality offers a huge inventory of
> difference, and ceaselessly criticises this trap; it has insisted
> particularly on the impossibility for the believer of stopping on
> the 'moment' of the break – a practice, a departure, a work, an
> ecstasy – and of identifying faith with a site.[35]

De Certeau actually uses monastic language to describe his
understanding of Christianity. Christianity, he says, is a 'way of
proceeding'.[36] This phrase is a direct allusion to the language of the
Constitutions (or rule) of the Jesuit Order to which he had
belonged. It has often been noted that much of the difficulty in
unravelling de Certeau is that he worked within such an eclectic
range of disciplines. Perhaps we have ignored an additional point
expressed in the story that the French Jesuit Provincial believed
that de Certeau certainly died a Jesuit but one couldn't be sure he
died a Christian! I believe that we can only really understand de
Certeau if we also understand his (loosely speaking) 'monastic'
background. In a way it was the monastic project that led him,
paradoxically, to monasticism's fundamental asceticism of leaving
structured, 'placed' forms to move once again to the margins where
the monastic way, after all, began.

Utopias in the end elude easy definition. In Christian spirituality
the worlds created by monastic rules, and the communities that
derive from them, have some utopian qualities. They invite us to
discern the signs of a place to which we do not yet belong but in
which we will belong. Such utopias are 'no place' because they
remind us that we are displaced at present and do not yet fully
inhabit place. But they are signs of future belonging. They are, in
space and time, a 'not here yet' that is beginning to take place.

Perfect languages

Interestingly, utopian thinking often included a search for the perfect language. Sometimes this was perfectly ordered philosophical language, on other occasions the language of paradise.[37] Christianity has often sought to use a universal, non-vernacular language, for example Latin, Church Slavonic, the language of the *Book of Common Prayer*. A key idea was the one-ness of language which somehow reversed Babel and its symbolic dispersion. Monastic language might be Latin, or it might be sign language, or it might be the language of silence. Interpretations of silence differ in classic spiritual and monastic texts. Some would appear to see silence as superior to conversation and not to attribute the value of *edificatio* to the latter. Sometimes silence is described as a matter of self-discipline or as a goal in itself. These attitudes tend to go along with an emphasis on the virtues of the *individual* monk in isolation rather than on the spirituality of a community as a whole or on ministry to others. Other texts, often those inspired by the Rule of St Augustine rather than by the Rule of St Benedict, understand silence in more relational terms as the prolegomenon of meaningful speech.[38]

In the end, the search for the 'perfect language' is not so much about vocabulary, syntax or grammar as about the *practice* of language. Thus, in one Augustinian community, a major text concerning teaching about silence understood it as an education in abstention from lazy or useless words.[39] The monastic tradition from its origins has witnessed to the truth that it is important to take care with words. Language without thought can be misunderstood and destructive. 'One of the old men said, "In the beginning, when we came together, we spoke to the good of souls, we advanced and ascended to heaven; now when we come together we fall into slander, and we drag one another to hell."'[40] A key to monastic spirituality is the centrality of discernment and this applies above all to speech. This enables the person to distinguish between language that is destructive and language that brings life. Words, most of all, only have value if they are the external expression of a life of integrity. An old man said, 'Spiritual work is essential, it is for this we have come to the desert. It is very hard to teach with the mouth that which one does not practise in the body.'[41] Silence, therefore, is not anti-social nor self-punishment but a necessary reticence in order to correct over-hasty or unproductive speech. The Irish *Regula Monachorum* of St Columbanus puts this rather quaintly in its chapter 'Of Silence':

> Justly will they be damned who would not say just things when they could, but preferred to say with garrulous loquacity what is evil, unjust, irreverent, empty, harmful, dubious, false, provocative, disparaging, base, fanciful, blasphemous, rude and tortuous.[42]

The Rule of St Benedict notes that vulgarity, gossip and talk leading to laughter are forbidden in *all places* within the enclosure (RSB 6:8).

Positively, monastic place is where the monk learns the language of the heart as the necessary prelude to speech.[43] Although the original Benedictines certainly did not contemplate becoming preservers of culture, the necessity for the monk to read meant that monasteries were on a different linguistic level. The *Rule of the Master* set the tone for later Rules by specifying that all monks under fifty must learn to read if they did not do so on entry.[44] Reading (two or three hours daily) was part of the monastic diet, as the reading of Scripture was a necessary preliminary to *meditatio* or the oral repetition of biblical verses committed to memory that was itself preliminary to *contemplatio* or the ultimate expression of single-hearted concentration on God. This meant that the monastery became a culturally privileged place with library and scriptorium.[45]

Common life

Utopias and monasteries suggest alternative visions of human community. Medieval theologians, who first speculated about the nature of the eternal bliss of the Kingdom, differed about the presence of community, social life and conversation as part of the package. Aquinas naturally followed orthodox doctrine in appreciating the place of the body in the eternal bliss of individuals. However, he was less happy to acknowledge social joy among the saints. He knew from Revelation that eternal life would not be spent in solitude with God but in the company of the angels and fellow saints. However, God must be the exclusive source of happiness and fulfilment. Thus 'fellowship of friends is not essential to happiness, since the human being has the entire fullness of perfection in God'. However, he also conceded that the bliss of the saints is supported 'by the fact that they see one another and rejoice at their fellowship in God'.[46] In contrast, Giles of Rome, a pupil of Aquinas, took a rather different view. The saints are a *societas perfecta* who retain a social life based on communication:

> When indeed it is argued that society depends upon language, then we have to say that in a state in which society is not abolished but made perfect, language, too, is not abolished but also made perfect . . . To be able to speak is not a sign of imperfection, but of perfection – and everything perfect must be said of the saints.[47]

Christian communities such as monasteries offered quite specific suggestions as to how real people can be converted into Kingdom people. They share with many utopian visions (for example, communism) a sense that most human divisions have to do with private property or indeed with privacy of any kind. The renunciation of personal property and money is a key element of monastic asceticism. This partly reflects a theological belief. The individualism that permeates so much of Western culture would have been alien to the world-view of early Christian ascetics. Augustine is typical in believing that it is humanity, rather than autonomous individuals, that is created in the image of God. Virtue consists of defending what is held in common. There will be no room in the Kingdom of God for a self-enclosed and protected privacy.[48]

What is interesting in monastic utopias is that manual labour was given a central role whereas in paradise work was unnecessary. So what is the meaning of work? Is it a reminder that monastic place, while utopian was only so proleptically? Humanity still has to live by the sweat of the brow and the illusion of the leisured classes was that some people could buy their way past this burden! Or did work stand for a radical social equality? Probably a mixture of both. Almost all monastic reforms linked the restoration of communal living as against the adoption of private space and private property to the restoration of manual labour to a central position. Interestingly, John Milton's *Paradise Lost*, unlike Genesis, has Adam and Eve working in the Garden. This was their response to prodigious fruitfulness. There are elements of this in monasticism as well, so that work in monastic places cannot simply be seen as a continuation of the human curse of sin.

The late-fourth-century *Ordo Monasterii*, part of the Rule of St Augustine, addressed common life in these terms: 'No one is to claim anything as his own, whether clothing or whatever else; we wish to live as the apostles did.'[49] In the *Praeceptum*, or Rule proper, the common life is equally stressed as is the imitation of the community in Acts. However, there is also space for individual differences and needs:[50]

> 2. In this way, let no one work for himself alone, but all your work shall be for the common purpose, done with greater zeal and more concentrated effort than if each one worked for his private purpose. The Scriptures tell us: 'Love is not self-seeking.' We understand this to mean: the common good takes precedence over the individual good, the individual good yields to the common good.[51]

Material possessions are rejected in the Rule of St Benedict:

> If he has any possessions, he should either give them to the poor beforehand, or make a formal donation of them to the monastery, without keeping back a single thing for himself, well aware that from that day he will not have even his own body at his disposal. Then and there in the oratory, he is to be stripped of everything of his own that he is wearing and clothed in what belongs to the monastery. (RSB 58:24–26)[52]

Celibacy is another important 'boundary' of monastic utopias. Along with property, families were seen since Plato's *Republic* as the most frequent causes of favouritism and corruption. Celibacy was not simply a rejection of sex or indicative of the utter sinfulness of the body. It indicated a commitment to the Kingdom in the sense that it was believed that conventional partnership would not be a characteristic of the *societas perfecta* of the Kingdom.[53] Celibacy symbolized singleness of heart. It also removed monastic people from a complex of social and economic relationships. This created a kind of liminality, a removal to the margins, that nevertheless had a social value. For women, celibacy had another social meaning. It offered them an equal partnership with men as disciples of Christ. It offered women a spiritual power and authority that was normally reserved for men in the 'natural world'. In that sense, celibacy could stand for a different kind of society in a different kind of world.[54]

Stability

The monastic vow of stability embeds the imaginative world firmly in a commitment to actual places. The vow of stability commits each person to seeking the Kingdom in *this* place and no other, by *remaining*. It mimics the radical commitment of God to the human condition. Yet monastic place also stands for a belief that through stability in a physical place boundaries may be crossed and the critical *inner* journey undertaken. The monastery is not so much a place set apart as the place of 'the other', the 'imagined', the 'desired', in the midst of everyday life.

The desert tradition of monastic life placed a central emphasis on the importance of staying in one place, specifically the 'cell', in order to find God. The point was stressed that if one could not find God in stability, there was no guarantee that God could be experienced by moving anywhere else. Stability in itself was also a protection against seeking to assuage spiritual boredom by wandering hither and thither.

The principle key to stability was to remain in the cell:

An old man said, 'Just as a tree cannot bring forth fruit if it is always being transplanted, so the monk who is always going from one place to another is not able to bring forth virtue.'

An old man said, 'The monk's cell is like the furnace of Babylon where the three children found the Son of God, and it is like the pillar of cloud where God spoke with Moses.'

The Fathers used to say, 'If a temptation comes to you in the place where you live, do not leave the place at the time of tempta-tion, for wherever you go you will find that which you fled from there before you.'

A brother asked an old man, 'What shall I do? For my thoughts trouble me, saying, "You can neither fast nor work, at least go and visit the sick, for this also is charity."' The old man said to him, 'Go, eat, drink, sleep, only do not leave your cell, for you must realise that it is endurance in the cell that leads the monk to his full stature.'[55]

'The workshop where we are to toil faithfully at all these tasks is the enclosure of the monastery and stability in the community' (RSB 4:78). Monks are to remain in the physical enclosure except for necessary journeys (RSB 50–51). Also (Prologue 50) the monk is to 'observe [Christ's] teaching in the monastery until death'. The point, however, is that stability of place only makes sense as a con-text for stability of intention and commitment.

Space of reconciliation

Monasticism is not only a metaphorical map of reconciled space. Boundaries are important to monastic space, not only as markers of exclusion but also as a way of shaping what is within into a particu-lar kind of place. It was not easy to enter:

> Do not grant newcomers to the monastic life an easy entry, but, as the Apostle says, 'Test the spirits to see if they are from God' (1 John 4:1). Therefore, if someone comes and keeps knocking at the door, and if at the end of four or five days he has shown him-self patient in bearing his harsh treatment and difficulty of entry, and has persisted in his request, then he should be allowed to enter and stay in the guest quarters for a few days. (RSB 58:1–4)

Monastic space was a theological space that reconciled love and knowledge. Patristic and, later in the West, monastic theology pre-supposed a seamless robe of intellectual reflection, prayer and praxis. Knowledge of divine things was inseparable from the love of God deepened in prayer. As in Augustine (*On the Holy Trinity*, Books XII–XIV), God is known not by *scientia* but by *sapientia* – that is, not by analysis but by a contemplative knowledge related to love and desire. Such an approach to theology had a context. The move towards theology as intellectual speculation during the twelfth and thirteenth centuries was associated with a change of context away from monastic space to intellectual space in, first, the cathedral schools and then the new universities. The space of theology became disassociated from the space of contemplation and *ascesis*.

Unlike the Rule of St Benedict, the Rule of St Augustine offers a quite prolonged treatment of how monastic space is one of recon-ciliation with 'the other', particularly in reference to crossing social boundaries:

> 6. Nor should they put their nose in the air because they associ-ate with people they did not dare approach in the world. Instead they should lift up their heart, and not pursue hollow worldly concerns . . .

> 7. But on the other hand, those who enjoyed some measure of worldly success ought not to belittle their brothers who come to this holy society from a condition of poverty. They should endeavour to boast about the fellowship of poor brothers, rather than the social standing of rich relations . . .[56]

The boundaries of religious enclosures also marked off a realm where spiritual powers, or the power of good, predominated over evil. Within the enclosure the rules of raw nature gave way to the sacred powers of ritual and prayer. Beyond the boundary wall lay a 'wilderness' that was not only naturally uncultivated and socially untamed but also the dwelling-place of demons and the forces of evil.[57] In Celtic monastic settlements the boundary further marked out a legal area that was regarded as sacred. This enclosure, or *termon*, was to be a place free from all aggression. Violence was legally and absolutely excluded from this precinct. This was a specifically Celtic version of the more general acceptance throughout Christendom that monastic settlements were places of sanctuary. Because of this absence of violence, monastic and other religious settlements also functioned as a kind of bank in which valuables could be deposited for safekeeping. Some of them also served as the equivalent of an open prison. Large settlements had groups of penitents attached to them, some of whom had committed serious crimes against society, such as murder.[58]

Monks in the desert tradition saw monastic settlements as anticipations of paradise in which the forces of division, violence and evil were excluded. In moving away from the *polis* in favour of the *eremos,* the ascetic was not in some simple way rejecting culture for nature – after all, the desert signified wildness, danger and suffering rather than beauty or romantic harmony. Rather, the monastic ascetic sought a third way – a reconfiguration of disordered place into the restoration of a prelapsarian paradise that was, at the same time, an anticipation of the final restoration hoped for with the incoming Kingdom. Wild beasts were tamed and nature was regulated. The privileges of Adam and Eve in Eden, received from God but lost by the Fall, were reclaimed. Many utopias stress that humanity should live in harmony with nature. To live in harmony with one's *own* nature was, of course, a major motive behind monastic teachings on asceticism. The latter was not so much an unbalanced rejection of aspects of human life as a disciplined attempt to keep human life in proper balance. The harmony with the wider world probably explains the many legendary tales of ascetics living in friendly proximity to wild beasts. This has sometimes been seen in terms of the virtue of the saints. In fact it has as much to do with the conversion of the *animals,* whose natural tendency would not be to be friends of humans. Within the enclosure, all wildness was tamed. Indeed, monastic life was frequently seen in terms of taming the wilderness.[59]

Receiving the stranger

All guests who present themselves are to be welcomed as Christ, for he himself will say: 'I was a stranger and you welcomed me.' Proper honour must be shown to all, especially to those who share our faith and to pilgrims.

Great care and concern are to be shown in receiving poor people and pilgrims, because in them more particularly Christ is received; our very awe of the rich guarantees them special respect. (RSB 53:1–2, 15)

A central aspect of monasteries as places of reconciliation concerns the reception of strangers.[60] Receiving the stranger as if they were Christ was an evangelical mandate for all Christians (Matt. 25.35) and thus became integral to monastic identity as a microcosm of the Christian life. All the major figures of early monasticism such as Basil of Caesarea, John Chrysostom and Jerome founded specialized institutions for hospitality including pilgrim hospices and hospitals. The collections of sayings and anecdotes from the fourth-century and fifth-century Egyptian desert fathers and mothers are full of examples of hospitality as a rule of life. Charity was an essential element of the search for singleness of heart:

A brother went to see an anchorite and as he was leaving said to him, 'Forgive me, abba, for having taken you away from your rule.' But the other answered him, 'My rule is to refresh you and send you away in peace.'

A brother questioned an old man, saying, 'Here are two brothers. One of them leads a solitary life for six days a week, giving himself much pain, and the other serves the sick. Whose work does God accept with the greater favour?' The old man said, 'Even if the one who withdraws for six days were to hang himself up by the nostrils, he could not equal the one who serves the sick.'[61]

The Rule of St Benedict (chapters 53 and 61) gave a central place to hospitality while protecting other aspects of the life from disturbance. Hospitality was personal and face-to-face, though surrounded by ritual and carefully circumscribed interaction.

Hospitality is always a blending of inside and outside. In other words, hospitality creates a 'between' place. This is where the other is encountered and social difference transcended. This is a critical element in a spirituality of reconciliation. Monastic space is marginal both in the sense that it exists on the edge of society and in the sense that it exists at a perceived intersection of local place and transcendent place. Sometimes the liminality of hospitality is expressed structurally as, for example, in some Irish monastic settlements. Passing guests were accorded a kind of semi-spiritual status and housed within the sacred enclosure. Often the guest house was given the choicest site in the settlement and yet was always set apart, sometimes within its own enclosure. The *hospitium*, therefore, was within the sacred space (isolated from the outside world) yet separated from the monastic living quarters. The guest quarters was itself, therefore, a kind of 'boundary place' between two worlds.[62]

Anti-domesticity

I have mentioned that 'home' is a preoccupation of some contemporary philosophical and anthropological approaches to place. Paradoxically, there is considerable use of domestic imagery in monastic writings but there is also a pervasive 'religious rejection of home'.[63] The Kingdom of God and this-worldly utopian anticipations of the Kingdom replace normal home as the focus of religious life. Our current domestic ideals are de-emphasized in favour of the asceticism of exile. This anti-domesticity was sanctioned by New Testament texts especially those in the Lucan tradition (e.g. Luke 14.26–27). Interestingly, monasticism developed strands that emphasized displacement and exile in a quite literal way (for example, Celtic forms) and also a more metaphorical emphasis on pilgrimage. This is a key image in Augustine:

> What makes a Christian's heart heavy? He is a pilgrim and yearns for his country. If for this reason your heart is heavy, although you enjoy worldly prosperity, you yet groan. And if all things work together to make you fortunate, and the world in

every way smiles on you, you nevertheless groan because you see that you are set on a pilgrimage. You feel that in the eyes of the foolish you certainly possess happiness but not as yet according to Christ's promises. You seek this with groans, you seek it by longing. By longing you ascend, and while you ascend you sing the Song of Steps, saying, 'To you I lift up my eyes, O You who dwell in heaven.'[64]

Monasticism stands for the fact that the Christian vision of true humanness involves a transcendental ideal. Christianity is nostalgic for a home where we have never been. Yet, Christianity rejects the notion that ordinary experience is an illusion. But if what we seek is somehow related to ordinary experience why do we not know it? It is because the primal ground of our experience is confusion, fragmentation or selective attention. In that sense, Christianity points to a transcendence that has always been here but which we do not notice. Monasticism does not suggest that the ordinary is irrelevant but that our perception of what truly is, is undermined by inattention. The very word *monachos*, from the Syriac word *ihidaya*, is best translated as 'the single one'. This singleness has a number of dimensions, but here the critical one is that the monk is the one who is 'single minded'.[65] This is why one of the greatest faults in the minds of the monastic way has been a divided heart and the greatest virtue is singleness of heart. The monk was 'a man who had gained a heart that was all of one piece, a heart as unriven by the knotted grain of private, unshared meanings and of private, covert intentions as was the solid, milk-white heart of the date-palm'.[66]

Journey: place as universal

Finally, we should recall once again that there is a persistent tension in Christianity between what I called the local or particular and the universal or 'catholic' dimensions of place. This tension exists in the monastic way especially in terms of stability, yet of life as a journey. Columbanus (*c.* 543–615), the great Celtic monastic founder across much of Western Europe, employed 'road' and 'journey' as his favoured metaphors for the Christian life. Life itself was a roadway that led to eternity:

> Let us concern ourselves with things divine, and as pilgrims ever sigh for and desire our homeland; for the end of the road is ever the object of travellers' hopes and desires, and thus, since we are travellers and pilgrims in the world, let us ever ponder on the end of the road, that is of our life, for the end of our roadway is our home.[67]

The old Irish life of Columba also suggests that pilgrimage is of the essence of Christianity rather than merely an eccentric practice of some ascetics:

> God counselled Abraham to leave his own country and go in pilgrimage into the land which God has shown him, to wit the 'Land of Promise' . . . Now the good counsel which God enjoined here on the father of the faithful is incumbent on all the faithful; that is to leave their country and their land, their wealth and their worldly delight for the sake of the Lord of the Elements, and go in perfect pilgrimage in imitation of him.[68]

The pilgrim was a *hospes mundi*, a guest of the world. Columbanus preached the essential instability and transitory nature of earthly life. Life is a roadway which the Christian must travel in perpetual pilgrimage. 'Therefore let this principle abide with us, that on the road we so live as travellers, as pilgrims, as guests of the world [*ut hospites mundi*].'[69]

Here there is a deep connection with the monastic ideal of 'spiritual poverty' and the rejection of material possessions as the measure of personal identity. The wanderers witnessed to the radical equality of all people before God. They made themselves displaced persons. God alone became their country of origin and faithfulness to God was what gave them status. This was powerfully expressed by travelling not only to where they were not known but to where they did not even understand the language.[70]

Although by the ninth century the Irish tradition of wandering monasticism was declining in the face of increasing criticism, it re-emerged in a new form during the thirteenth century in the birth of the mendicant movement of new religious communities. Already the dominant monastic spirituality of stability and separation from the world was breaking open into broader spiritual unsettlement of crusade and pilgrimage. Francis of Assisi embraced this instability of life on the road in his *Later Rule*. One phrase in particular expresses an understanding of discipleship that accords with the evangelical and penitential movements of his age. 'As pilgrims and strangers in this world who serve the Lord in poverty and humility, let them go begging for alms with full trust.'[71]

Within the monastic way, broadly understood, one of the most striking examples of the emphasis on place as universal, catholic and unbounded is the Rule of St Ignatius Loyola, written in the middle of the sixteenth century. The *Formula of the Institute* and the *Constitutions of the Society of Jesus* eventually became the most influential post-Reformation Rule in the Western Church.[72] We know that Ignatius Loyola and his Secretary, Juan de Polanco, assiduously researched previous monastic Rules including the Rule of St Basil and the so-called Rule of Pachomius (from its Latin translation as the Rule of St Jerome), the writings of Cassian, the Rule of St Benedict, the Rule of St Augustine, the Rule of St Francis and the constitutions of the Order of Preachers.[73] That said, the Jesuit 'rule' departed from traditional practice in having no obligatory liturgy in common, no distinctive dress, few visible practices that distinguished members from other reformed clergy and no system of capitular government. All this favoured a spirituality of mission and mobility that contrasted strongly with the traditional monastic emphasis on conventual life and common prayer.[74] The fully professed members took a fourth vow to journey anywhere in the world to undertake ministry. Although this was described as a vow of special obedience to the pope, its specific content was to travel:

This is a vow to go anywhere His Holiness will order, whether among the faithful or the infidels, without pleading an excuse and without requesting any expenses for the journey, for the sake of matters pertaining to the worship of God and the welfare of the Christian religion.[75]

At the beginning the members assumed that most Jesuits would not reside for very long in the professed houses but would spend most of their lives in journeying. 'One should attend to the first characteristic of our Institute . . . this is to travel' (*Const.* 626). Ignatius Loyola's other principal assistant, Jeronimo Nadal, wrote a great deal by way of commentary on the deeper meaning of the Fourth Vow of the professed. Thus the 'best house' of the Jesuits was their journeying for ministry:

> The principal and most characteristic dwelling for Jesuits is not the professed houses, but in journeyings.

> They consider that they are in their most peaceful and pleasant house when they are constantly on the move, when they travel throughout the earth, when they have no place to call their own.[76]

Again, in the words of Nadal, 'The world is our house'. Jesuit place is the *oikumene*, the whole inhabited world. This was not merely a statement about mobility and the geographical scope of Jesuit life. It has deeper theological resonances with a reconciliatory theology of nature and grace. It is not 'monastic space' alone that is a place of reconciliation but the *space of the world*. This view permeated not only the writings of Nadal but also the *Constitutions*.[77] This benign theology of a graced world, albeit one marred by the sinfulness of human actions, is a major theme of the spirituality of the *Spiritual Exercises* of which the Jesuit *Constitutions* were one concrete expression. In the famous contemplation on the Incarnation, the world, the whole of creation and of human history are to be seen only through the eyes of the Trinity who desire the healing and reconciliation of all things.[78]

In this sense, what had been a closed and protected utopian vision of monastic enclosure in other spiritual traditions is broken open in the Jesuit *Constitutions* and in the availability of the spirituality of the *Exercises* to all manner and condition of women and men in their own everyday places.

5

The Mystical Way: Transcending Places of Limit

I want to suggest that mysticism also concerns our 'practice of place' and to ask whether this concerns merely a private and protected world of interior space. Whatever we mean by the mystical dimension of Christian faith and practice, I suggest that it both subverts our temptation to settlement and impels us into a condition of perpetual departure. I should add that the issues I will address are largely specific to the Western tradition of theology and spirituality. The Eastern tradition, in its understanding of mystical theology, has continued to develop within what may be thought of as an integrated patristic framework.

The problem of definition

As we examine the concept of mysticism, we immediately face a problem of definition. Despite the large body of modern reflection on the subject, mysticism is notoriously controversial and difficult to define. Most writings until relatively recently have concentrated on mysticism as a discrete category of spiritual experience. This approach results in three problems. First, it tends to separate mysticism from theology. This leaves mysticism with no relevance to the ways we attempt to think or speak about God, and relegates it to a devotional backwater. Second, it removes mysticism from the public world into private, individual, interior space. The result is that it is difficult to suggest how mysticism may be of any great importance ecclesially or socially. Third, it tends to concentrate on phenomena or 'states of mind' and emotions experienced by a limited number of rather eccentric people as the result of intense meditative practice. This effectively separates mysticism from Christian discipleship in general. Such approaches also raise the question as to whether, given that a mystical dimension exists in all world religions, there may be a something, 'mysticism as such', that transcends the boundaries of particular religions and which is susceptible to generic definition. This is a common viewpoint in many modern popular treatments of mysticism, especially those influenced by the so-called New Age with its innate suspicion of closed intellectual or dogmatic systems. However, in practice all attempts at providing inclusive definitions of mysticism are open to the criticism that they are thoroughly inadequate in relation to the history, practices and perspectives of specific religious and cultural traditions. Further, are we to limit the term 'mystic' to people who speak of first-hand experiences (especially ecstatic or visionary ones), such as Julian of Norwich, or include writers whose style is apparently more detached and theoretical, for example, Bernard of Clairvaux?

There have been recent attempts, for example in the monumental history of Western mysticism by Bernard McGinn (still in progress), to move away from a concentration on experience towards defining mysticism in terms of the immediacy of the presence of God. I am not clear how far this actually changes the fundamental direction. It still appears to focus on externals rather than on the core of the matter. Far more valuable is McGinn's astute comment that, whatever we mean by mystics, such people do not set out to practise mysticism but simply to live their Christian discipleship and faith at depth.[1] Mysticism, therefore, becomes a dimension of discipleship or a word we use to express the process of the growing intensification of Christian life and practice, accessible to all rather than to any elite.

In recent decades a number of major theologians such as Karl Rahner or Rowan Williams have re-engaged with the concept of mysticism, or better the mystical dimension of theology, in terms of a different way of knowing and learning beyond a purely intellectualist, abstract or systematic method.[2] Because this way of knowing necessarily takes us beyond the boundaries of conceptual thinking and beyond definitive intelligibility, there has been a particular interest in the recovery of the apophatic dimension of mystical theology with its emphasis on the final impossibility of naming God as this or that. Williams understands this approach as ultimately normative within the overall theological enterprise. 'Apophasis is not a branch of theology, but an attitude which should undergird *all* theological discourse, and lead it towards the silence of contemplation and communion.'[3] This view is reinforced by the journey of the American theologian, David Tracy, towards a belief that the apophatic language of mystical writers is where theologians must turn in the present, postmodern era. He writes of the 'uncanny negations' of mysticism as a form of theological release and, consciously aligning himself with the later Thomas Aquinas, that intellectual silence may be 'the final form of speech possible to any authentic speaker'.[4]

The basis of Christian mysticism

Where does the notion arise that there are such things as mysticism, mystical experience, or mystics, in the sense of a definable set of practices or experiences, confined to certain uncommon people who are, in some spiritual sense, privileged? Mysticism in this sense is probably an invention of nineteenth-century scholarship, particularly William James.[5] Consequently, when we examine the earlier tradition, it would be wrong to see the use of the metaphors of inwardness or ascent as simply implying the achievement of a certain kind of individualistic experience.

In its origins during the patristic period, the concept of mystical theology had nothing to do with the later preoccupation, especially in the West, with subjective inward experience. Patristic mysticism refers to the collective lives of Christians who come to know God in Christ by incorporation into the fellowship of the 'mystery'. The community of believers is drawn ever deeper into the mystery of God through exposure to the Scriptures and participation in the sacred liturgy. So, mysticism in its fundamental sense is 'to live the mystery' of incorporation into Christ beginning with baptism and fed by the Eucharist.[6] This liturgical and public understanding of 'the mystical' found its classic expression in the writings of the early-sixth-century anonymous author commonly known as Pseudo-Dionysius. He had a profound impact on the development of Western spirituality during the Middle Ages.

This story of origin of Christian mysticism underlines the fact that mysticism, in its most general sense, is potentially a dimension of every Christian's life. It also emphasizes that mysticism is a baptismal *gift* rather than something achieved through individual effort. In our contemporary world we may wish to add to Scripture and liturgy, as the media of mystical deepening, another dimension of discipleship – engagement with the task of transforming the world and with the structures of human living. I will return to this later. Certainly, the traditional phenomena of mystical experiences are not reliable indicators of the mystical dimension of discipleship. Mysticism, if we continue to use the word, is essentially a process or way of life rather than a set of isolated experiences. It is, indeed, a way of speaking of the Christian *life* lived with particular commitment and intensity.

Even the later Western medieval sense of mysticism, while it detached the theology of Pseudo-Dionysius from its community and liturgical roots to some degree, does not fit fully with an emphasis on protected interior space. This approach owes far more to a modern preoccupation with affectivity and 'experience'. There is a danger that mysticism in its modern guise becomes another form of 'positivist' thinking that takes over from dogmatics as a new framework of certainty that is immediate, foundational, free from theoretical presuppositions, and capable of verifying religious truth.[7] This merely relocates certainty from external institutional structures to 'the inner self'.

Contemporary Western culture encourages us to be preoccupied with our identities as individual selves. However, in the mystical tradition, what I am in the last resort is what I am in my deepest interiority. However, in this deepest interiority (for example in Julian of Norwich, based on Augustine), God and the human person meet in a union beyond description and beyond experience. It is not only God who is beyond description but also my 'self' which, according to Julian, for example, I can only come to understand the deeper I am plunged into the mystery of God.[8]

Some contemporary Western intellectual developments, especially what is called postmodernism, point to the possibilities for a revival of medieval apophatic deconstruction. What is critical, however, is that medieval apophatic theology was not merely an intellectual critique of positivist discourse but a *practice* embodied in a way of living within the community of the Church and in public society. The Christian mystical tradition suggests not a private world of inner experience but a consciously organized 'strategy of disarrangement as a way of life, as being that in which alone God is to be found'.[9] The negativity of the apophatic tradition does not consist merely of a consciousness of the absence of God. That is scarcely different from a consciousness of God as present. It would be more appropriate to speak of the absence of experience. 'Is it not better to say, as expressive of the apophatic, simply that God is what is on the other side of anything at all we can be conscious of, whether of its presence or of its absence?'[10]

It is critical that the language of exteriority or interiority, of withdrawal and engagement, affirmation or denial, be placed in a dialectical relationship.[11] The path of spiritual progress is often described in terms of interiority. So, following Augustine, the path must lead progressively from a life lived in 'exteriority' or activity to an 'interior' or contemplative life. But at another level, the language of interiority actually overcomes such a distinction. The truly interior person, by becoming more and more deeply drawn into the Trinitarian mystery of God, is able to transcend such categories of existence as 'exterior' and 'interior'. So such a person lives 'not "within" but in a "nowhere" which is an "everywhere"'.[12] Some of the more interesting later writings of the twentieth-century Cistercian Thomas Merton spell this out in terms of a growth of universal compassion and an ever increasing and painful sensitivity to the world and to all elements of human existence.[13]

Excursus on visions

On this basis, it is possible to reject out of hand the significance of ecstatic experiences and visions as no more than eccentric states of mind – perhaps artificially induced – or even indications of psychological instability. The argument is that this emphasis removes the mystical dimension of religion beyond the life of ordinary believers into a narrowly defined and individualistic track. Unless we are to say that the subjective experience of certain phenomena and the achievement of certain psychological states *is* the purpose and justification of mysticism (associated with the practice of certain meditative techniques), the major questions concerning the overall role or value of mysticism within a community of faith are left unanswered. The emphasis on special visionary experiences, indeed 'experiences' of any kind, also reduces the focus for the study of mysticism, particularly Christian mysticism, merely to those texts that fit within the genre of first-hand accounts. Conversely, such an emphasis might suggest that all such extraordinary phenomena should be categorized as 'mystical' and those who experience them as 'mystics'.[14]

Having said this, I believe that there are also dangers in hastily dismissing descriptions of visionary experiences as irrelevant epiphenomena which some people may or may not have as a by-product of intensive contemplative practice. For example, a great deal of detailed study has been done on the mystical writings of Cistercian nuns at Helfta during the thirteenth century. This community had close associations with Beguine mysticism.[15] It has been noted that the tremendous growth in visionary experiences was especially prevalent among groups who did not have other sources of religious authority. Visions opened up a quasi-public space for teaching roles. Gathered around the nuns at Helfta there was a wide circle of male and female admirers. Those whose mystical visions were recorded were seen not merely as intercessors but also as authorised in some way as preachers or teachers.

The 'content' of the visions sometimes explicitly included references to priestly garb or sacramental powers. This was a period of greater clericalization in the Western Church and the curtailing of the ability of women to exercise spiritual ministry in the public forum.[16] However, while visionary experience offered some kind of authorization aside from the power of office, we must be careful about overestimating the degree to which this strand of spirituality subverted the world of male public authority, whether secular or religious. Clearly, because the nuns (and other female visionaries such as certain Beguines or Julian of Norwich) were able to enhance the status of female teaching in an unfavourable climate, there was a subtly subversive quality to their writings. In the language of Michel Foucault, the discourse of female mystics may be thought of as a kind of heterological space. In terms of de Certeau's analytical framework, such 'places' were a kind of segregated 'refuge', a circumscribed space for a practice of meaning that represented a rupture with the public, clericalized space of the Church. On the other hand, the American scholar Caroline Walker Bynum, for example, notes that the attendant spirituality was generally deeply orthodox and actually enhanced the authority of the priesthood.

Although Grace Jantzen in her recent writings on gender and mysticism is also profoundly suspicious of defining a general phenomenon known as 'mystical experience', she departs from the work of more 'gender blind' commentators in directly addressing the issue of visions. It would be unreasonable to reject male approaches to mysticism just because they are male. However, we can no longer ignore as unimportant a tendency for male analyses of mysticism to accord more value to reason and to be suspicious of the sensory faculties commonly associated with the mysticism of women. It can be noted that, for Julian of Norwich and other female visionaries, visionary experiences not only endorsed their authority but also had a teaching value in that the imaginary world they conjured up offered access to a deeper level of understanding than was possible through a purely didactic style.[17]

Mysticism and social place

The excursus on visions brings me to the assertion that the mystical dimension of Christian practice has profound social implications. I find it disturbing that a number of contemporary post-Christian proponents of a new mysticism, and some Christian theologians who are critical of the contemporary drift to mysticism, continue to suggest that it is a 'tropical luxuriance' with no role in public, political places.[18] This view will simply not do from either a historical or a theological point of view. Christianity has persistently fought a theological and spiritual battle against the privatization of place. So-called mystical texts, properly understood in their contexts, do not support such a reductionist viewpoint. I suggest that the most substantial representatives of Western mysticism were opposed to rather than supporters of privatized experience. I will take two examples, both from the fourteenth century – the male Flemish cleric John Ruusbroec and the female lay visionary Julian of Norwich.

The Flemish theologian John Ruusbroec conceived of mysticism in terms of 'the life common to all'. This life joined created beings to each other in the service of all and harmonized the initially distinct moments of action and contemplation into a coincidence. So, the 'elevated' person is the 'common' person. The elevated person 'owes himself to all those who seek his help' and seeks to share the 'life common to all' that is of God:

> A person who has been sent down by God from these heights is full of truth and rich in all the virtues . . . He will therefore always flow forth to all who need him, for the living spring of the Holy Spirit is so rich that it can never be drained dry . . . He therefore leads a common life, for he is equally ready for contemplation or for action and is perfect in both.[19]

Ruusbroec understood the highest degree of the spiritual life to be expressed in a community-minded person who wished above all to share what he or she had received with everybody else. In that sense Ruusbroec relativized the purely contemplative life and rejected the common medieval interpretation (see *The Cloud of Unknowing*, for example)[20] of the Martha and Mary story in Luke 10. This was usually described in terms of a distinction between a lower, active life and a higher, contemplative life. In a life truly one with God, contemplative union and self-giving work alternate in a way that transgresses the boundaries between interiority and exteriority. This is a life that is truly at one with the rhythm of the inner life of the Holy Trinity:

> The Spirit of God exhales us so that we may love and awaken to virtue. He draws us back into Him to rest and to enjoy: that is eternal life, just like we exhale the air that is in us and inhale new air, because that is what our mortal life consists of.

> Even though our spirit is unspirited and its work falls away in enjoying blessedness, it is renewed afterwards in grace and love and virtue. To go into an inactive enjoying and to go out in good works and always to stay one with the Spirit of God, that is what I mean. As we open the eyes of our body, see, and then close them again, so quickly that we do not even notice, so do we die in God and live again out of God and remain one with God always.[21]

Ruusbroec saw Jesus Christ as the perfect example of 'the common person':

> We want to take Christ as an example, who was, is, and remains common [i.e. community-minded] in all eternity. He was sent to common humanity on earth, for the benefit of all men who would turn to Him . . . Observe now how Christ was common and gave Himself to all in true loyalty. His inner prayer was addressed to His Father and concerned all who wish to be saved. Christ was common in His love, His teaching and exhortation, and His comforting full of mercy and generosity. His soul and His body, His life and His death and His service were and are meant for all in common. His sacraments and His gifts are common to all. Christ never took food or drink for His sustenance without being aware of the general interest of all men who shall ever be saved down to the last day. Christ had nothing proper to Himself, nothing of His own, but everything was common: body and soul, mother and disciples, cloak and tunic. He ate and drank for our sake. Only His torments and His sufferings and His misery were proper to Him and His own, but the profit and the usefulness thereof are for the whole community and the glory of His merits will be eternally common to all.[22]

Ruusbroec was quite clear that those people who practise the attainment of a peaceful inwardness as the goal of their prayer and disregard charity or ethics are, of all people, most guilty of spiritual wickedness.[23]

Evelyn Underhill has been somewhat out of favour as an authority on mysticism. Yet her book *Mysticism* remains one of the classics of the last century.[24] One of her insights that continues to have value is that a defining characteristic of Christian mysticism is that union with the divine impels a person towards an active, outward, rather than purely passive, inward life. It seems to me that this viewpoint is associated to some degree with the views of Underhill's great mentor, Baron von Hügel, on the centrality of what he called 'inclusive mysticism'.[25]

> Hence, the ideal of the great contemplatives, the end of their long education, is to become 'modes of the Infinite'. Filled with an abounding sense of the Divine Life, of ultimate and adorable reality, sustaining and urging them on, they wish to communicate the revelation, the more abundant life, which they have received.[26]

This corresponds to my own perception over many years that the greatest contemplative figures in the Western spiritual tradition were also people of immense pastoral and prophetic energy. They moved naturally from a definitive placement in God to an engagement with the outer world to which that God was irrevocably committed.

In the use of spatial language, especially 'inwardness', contemplative practice as described by its most substantial commentators appears to suggest that experiences of inward and outward, near at hand and far away, are synthesized so that distinctions between them cease to have any meaning.[27] In Ruusbroec, for example, the consummation of spiritual love (the 'peace of the summits' as he calls it) involves an *active* meeting between the disciple and the Trinity in divine love and thus a dialectical relationship between resting and activity.[28]

> Understand, God comes to us incessantly, both with means and

without means; and He demands of us both action and fruition, in such a way that the action never hinders the fruition, nor the fruition the action, but they strengthen one another. And this is why the interior man lives his life according to these two ways; that is to say, in rest and in work. And in each of them he is wholly and undividedly; for he dwells wholly in God in virtue of his restful fruition and wholly in himself in virtue of his active love . . . Thus this man is just, and he goes towards God by inward love, in eternal work, and he goes in God by his fruitive inclination in eternal rest. And he dwells in God; and yet he goes out towards all creatures, in a spirit of love towards all things, in virtue and in works of righteousness. And this is the supreme summit of the inner life.[29]

For the great mystics such as Ruusbroec or Julian of Norwich, God is our country, our home and our place. Yet God is to be thought of as both in no place alone and yet in every place at once. To be drawn into the mystery of God implies, too, that ultimately the human person will find a home in a 'no place' that is at the same time a universal place.

Behind the distinctions of passivity and activity, interior and exterior, lies the question of the theology of God. The language surrounding the traditional belief in God's distance and detachment owes a great deal to the philosophy of Aristotle and particularly to the belief that perfection was a state of rest and to the concept of God as unmoved mover. This contrasts strongly with another tradition where God is understood as active in the place of the world. 'I will consider how God labours and works for me in all the creatures on the face of the earth; that is, he acts in the manner of one who is labouring.'[30]

What we refer to as mysticism is not a free-floating reality but always arises within specific contexts. In his original role as one of the pioneers of modern approaches to the history of spirituality, Michel de Certeau suggested that there were even 'privileged places' for the development of mystical insight including certain social categories. These were people with little or no power in a world of public places. De Certeau notes that in early modern

Europe these were 'social categories which were in socio-economic recession, disadvantaged by change, marginalised by progress, or destroyed by war . . . Aside from a few mystics on the road to social promotion . . . the majority of them . . . belonged to social milieux or "factions" in full retreat. Mysticism seems to emerge on beaches uncovered by the receding tide.'[31] Indeed, many were people who existed socially, culturally and even religiously on boundaries. A striking example noted by de Certeau is the Spanish mystical movement of the sixteenth century where an unusual proportion of the most significant figures came from the 'excluded' class of *conversos* or converted Jews. These include Teresa of Avila, Luis de Leon, Molinos and two figures central to the early Jesuits, Diego Lainez and Juan de Polanco.[32] Again, for someone like Grace Jantzen, 'mystics' and 'mysticism' are social constructions bound up with issues of power and power relations, not least between men and women.[33]

The emphasis on social, temporal and geographical context in relation to spirituality is now common in studies of mystical texts. This attention to context was ignored thirty years ago even by such otherwise sophisticated commentators as the brothers Karl and Hugo Rahner. For example, modern commentaries on Julian of Norwich all emphasize the particularity of England in the second half of the fourteenth century as one explanation not only for her *Showings* but also for the general proliferation of mystical literature, male as well as female. It was a period of immense instability. The Hundred Years War between England and France caused a continual death toll from every social class and a growing burden of taxation to pay for the war. At the same time there were persistent occurrences of plague. In Norwich a third of the population died in at least three major outbreaks between the late 1340s and the 1360s. A potent combination of war, taxation and plague created economic depression and social instability culminating in the Peasants' Revolt of 1381. There was also serious instability in the Church. Western Christendom suffered the Great Schism from 1378 to 1414 where kings, nobility, towns, universities, dioceses and religious orders were divided in their support for one of up to three papal claimants. In England there were the specific anti-clerical protests by the priest John Wyclif, his departure from Oxford in disgrace in 1382 and the growth of the popular Lollard movement which, loosely speaking, drew on some of Wyclif's ideas for reform. Indeed, some scholars draw attention to Julian's possible preoccupation in parts of her two texts with protecting herself against charges of heresy, whether Lollard or other.[34]

Mystics, especially but not exclusively women, were (indeed are) often confined in tightly enclosed or policed places. Contemplation then becomes a way of transcending such boundaries and of accessing a way of knowing that is beyond the power of authority to control. The small and concentrated space of contemplation, often within a monastic enclosure, becomes a place of expansion and a place to transgress normal boundaries.[35] Interestingly, Julian's mystical experience is described as taking place in the bedroom (personal, intimate, private place) but also in the presence of others (social space). It also involved a place of illness and dying and therefore of liminal space.

The remarks that follow reflect briefly on some elements of the mysticism of Julian of Norwich. They are extremely selective and do not cover the full range of ways in which Julian engages with the nature of place, whether particular or universal. In Julian we have a paradox. Hers was a restricted 'space' in so many ways. As an anchorite, she was physically separated from an external world in her hermitage. Yet, as we know from archaeology, such cells had windows on to two worlds. One overlooked the altar in the church alongside, the site of the ultimately 'catholic' place of the Eucharist. Another overlooked the street, the space of the world, where anchorites such as Julian could undertake a social and spiritual role as confidante and wise guide. She was bound in other ways by a male clerical culture. At the beginning of her Short Text she embraces the marginal cultural space:

But God forbid that you should say or assume that I am a teacher, for that is not and never was my intention; for I am a woman, ignorant, weak and frail. But I know that what I am saying I have received by the revelation of him who is the sovereign teacher.[36]

Although this disclaimer is omitted in the better known (and presumably later) Long Text, Julian nevertheless continues the same theme at the beginning of her Second Chapter. 'This revelation was made to a simple, unlettered creature'.[37] Without becoming embroiled in technical discussions about the meaning of 'unlettered', these sentiments coupled with the fact that the texts are in the Middle English vernacular and cite no theological sources make them a marginal place in relation to what we might call the theological and ecclesiastical 'centre'.

It is not insignificant that Julian's writings were in the vernacular. Indeed they are one of the most important and earliest English vernacular religious texts. This is more than a question of Julian's self-confession as one who is 'unlettered' – that is, presumably, unable to work in Latin because she came from a non-clerical world of experience. The development of vernacular during the fourteenth century was an important step in a change of religious sensibilities. As an appropriate vehicle for expressing mystical experience and teaching, the use of vernacular language was not simply a question of the democratization of the audience – that is, its extension to include lay Christians and women in particular. Language is not merely concerned with the *communication* of experience. It is associated with what kind of experience actually counted as truly spiritual. The growing acceptability of the vernacular during the fourteenth century as a medium for speaking of God and for spiritual teaching liberated everyday experience to become the place where God may be found.

Historical context is important. In the midst of her politically, economically, socially and religiously ruptured fourteenth-century world, Julian does not represent a refuge or a privatized space of interior spiritual delight accessible only to a mystic elite. Hers is a 'blessed teaching' written for others (LT 73). Her purpose is to emphasize a single revelation or showing, that of divine love, in order to liberate her 'even' (or fellow) Christians from all that prevents them growing into the life of God – especially the barriers of sin and despair (LT 73). Her teaching is for all who seek to love God, whatever their social position:

In all this I was greatly moved in love towards my fellow
Christians, that they might all see and know the same as I saw,
for I wished it to be a comfort to them, for all this vision was
shown for all men. (LT 8)

There are images of interiority in several Chapters of the Long
Text, not least in Chapter 55. There, God dwells enthroned in the
heart of the human soul. 'In that same . . . place exists the city of
God, ordained for him from without beginning. He comes into this
city and will never depart from it, for God is never out of the soul,
in which he will dwell blessedly without end.' On the other hand,
in Chapter 56 Our Lord Jesus is also said to dwell in human
sensuality which, in Julian's anthropology, is the changeable, falli-
ble, contingent dimension of human life and historical existence.
Further, God dwells in the heart of everything that is:

See I am God. See, I am in all things. See, I do all things. See, I
never remove my hands from my works, nor ever shall without
end. See, I guide all things to the end that I ordain them for,
before time began, with the same power and wisdom and love
with which I made them. (LT 11)

In other words, by placing Julian's portrayal of interiority within
the total context of the text of the *Showings* it becomes impossible
to defend the notion that it represents an abandonment of the world
of material place and of human history.

Julian is almost certainly explicitly dependent on Augustine's
theology of *vestigia Trinitatis*, the traces of the Trinity found particularly in the individual human soul. Certainly this relocation of
the 'economy' of God emphasizes interiority but not in such a way
as to undermine a connection between the Trinity and salvation. In
the very first 'showing' or revelation, Julian came to see that 'where
Jesus appears the blessed Trinity is understood, as I see it' (LT 4).
What is crucial in Julian is that the Passion of Jesus, and the
compassion of Jesus for humankind, is the measure of the life of
God-as-Trinity. Julian came to understand that everything she
was subsequently taught was grounded in the truth of that first
showing. The Passion is understood fundamentally as the supreme

revelation of the truth of God. This truth is universal compassion and love for the created order. 'All the Trinity worked in Christ's Passion, administering abundant virtues and plentiful grace to us by him' (LT 23).

In the Long Text, Chapter 68, Julian appears to adopt the language of ascent away from a world of places and history. The soul 'comes above all creatures into itself'. But this has to be taken in tandem with a much earlier passage where Julian explicitly rejects the metaphor of mystical ascent away from nature and the world of places in favour of a retreat to a purely spiritual zone:

> Then there came a suggestion, seemingly said in friendly manner, to my reason: Look up to heaven to his Father . . . I answered inwardly with all the power of my soul, and said: No, I cannot, for you are my heaven.

> So I was taught to choose Jesus for my heaven, whom I saw only in pain at that time. No other heaven was pleasing to me than Jesus, who will be my bliss when I am there. And this has always been a comfort to me, that I chose Jesus by his grace to be my heaven in all this time of suffering and of sorrow. And that has taught me that I should always do so, to choose only Jesus to be my heaven, in well-being and in woe. (LT 19)

Equally, the key to Julian's understanding of the 'ascent above' lies in her teaching about the importance of the little things of the created order and the importance of not settling finally in what is less than everything. In the famous passage of Chapter 5 of the Long Text, three important things emerge. First, the whole of reality is shown to Julian as something of extraordinary smallness and fragility:

> And in this he showed me something small, no bigger than a hazelnut, lying in the palm of my hand, as it seemed to me, and it was as round as a ball. I looked at it with the eye of my understanding and thought: What can this be? And I was given this general answer: It is everything which is made. (LT 5)

Second, in response to Julian's wonder that something so little should not have 'fallen into nothing', she is brought to understand that this is because God made it, God loves it, God preserves it. Yet, third, having established the infinite value of the space of creation, Julian also affirms, 'We need to have knowledge of this [the littleness of everything] so that we may delight in despising as nothing everything created, so as to love and have uncreated God.' Not only is the soul to despise everything created as nothing, it is to become nothing.

However, to treat all as nothing, to be made nothing, to be 'noughted', is not to deny the infinite value of the external world or to lose one's own identity. This 'noughting' can only be understood in terms of the ultimate point of the 'showings' which is to be free to receive the great gift God will offer at the end of all things – to make all things well (LT 32). The illusion that we may find ultimate satisfaction in what is partial, whether that is the outer world or a person's own self, has to be shattered so that we may enter, liberated, into what is whole and complete. 'For this is the reason why our hearts and souls are not in perfect ease, because here we seek rest in this thing which is so little, in which there is no rest, and we do not know our God who is almighty, all wise and all good, for he is true rest' (LT 5). And again, 'When it [the soul] by its will has become nothing for love, to have him who is everything, then is it able to receive spiritual rest.'

It is interesting that, in reviewing Julian's mysticism as a form of 'political theology', in the sense that it re-imagines the nature of the human *polis*, one recent study does not significantly engage with a highly relevant issue – the meditative genres with which Julian was likely to have been familiar.[38] The teaching of her book is based on what Julian refers to as sixteen 'showings'. I do not wish to become enmeshed in a rather sterile debate about the reality or otherwise of mystical 'visions', let alone in the complex classification of higher and lower forms of 'sight' (ghostly or bodily, for example) and 'locution' that preoccupied earlier Roman Catholic scholars.[39] What does not seem to be a matter of dispute is that Julian was in some way influenced by Franciscan forms of meditative devotion to the Passion of Christ.[40] This is not to suggest that the 'showings' were no more than descriptions of exercises in imaginative medita-

tion in the Franciscan style. However, the likely threads connecting
this meditative-contemplative tradition and the experience of the
'showings' have been commented upon.[41]

The term that is sometimes used to describe this meditative
form that emerged in the thirteenth century is 'a mysticism of the
historical event'. By this is meant that the particular style of meditation is focused firmly on the historical events of the life of Christ,
and especially on the Passion. The imagination is consciously
activated to visualize the gospel scenes. The meditater does not
view past events merely as past but imaginatively becomes a participant in the action in the present moment. The meditative
form involved entering the events not purely imaginatively but also
emotionally, intellectually – indeed, with the whole personality. By
this active involvement, the person meditating would find that the
events opened up their existential meaning and value. Detailed
studies of the tradition make it clear that the place of the meditative
or contemplative experience was not merely the past historical
events of Jesus whose energy is, as it were, unlocked for the contemplative. It is also the existential place which the person meditating inhabits in real space and time. The person is not asked to rise
above this context but to enter into it with ever deeper degrees of
perception and commitment.[42]

In terms of what I am suggesting about the visionary foundations
of Julian's teaching, the critical point is that the meditative tradition concerns far more than particular techniques or imaginative
flights of fancy. It concerns the historicity of human existence and
tends to give rise, for example in Julian, to a particular category of
mystical consciousness – one that creates a sense of unity with
the world of human places and history. This tradition acted as a
counterpoint to some other Western medieval contemplative traditions, influenced by elements of Neoplatonism, that tended to
project spirituality out of the world and human history into a time-
less elsewhere or 'no place'.[43]

As the relationship between context and content in Julian's texts
underlines, mystical texts are associated with attempts, at different
times and places, to respond to the needs of the moment. This
seems to apply particularly to moments of crisis, cultural break-

down or significant changes in human history. This perspective counteracts previous tendencies to view mysticism as detached from historical context, individualistic and concerned with private experiences of interiority. Such a view

Keeps God (and women) safely out of politics and the public realm; it allows mysticism to flourish as a secret inner life, while those who nurture such an inner life can generally be counted on to prop up rather than challenge the status quo of their workplaces, their gender roles, and the political systems by which they are governed, since their anxieties and angers will be allayed in the privacy of their own heart's search for peace and tranquillity.[44]

The contemplative practice that produced the texts we call 'mystical', despite all the possible criticisms of that term, is not essentially a privatized search for inner consolation but is an important aspect of a collective struggle for transformation.

In a much more specific sense, are there ways in which mysticism can be thought of as concerned with the place of politics? Echoing de Certeau, David Tracy suggests that mystics, like the mad, represent a kind of 'otherness' on the social margins. This 'otherness' has the capacity to challenge traditional centres of power and privilege.[45] Because the 'knowing' of mystical texts is predicated on union with God rather than on the power of human intellects to control reality, it bears some resemblance to the 'subjugated knowledges' spoken of by Michel Foucault. This resists dominant structures of power and knowledge and opposes established forms of discourse rather than simply offering a pleasing alternative.[46]

Within the tradition of liberation theology, the Chilean author Segundo Galilea has written more than almost anyone else concerning the mystical or the contemplative dimensions of political and social responses to injustice. Galilea suggests that there needs to be a movement from the notion that a political response is purely ethical or structural (which may merely become a new form of oppression) towards the truly spiritual experience of discovering the compassion of God incarnate in the poor. Humans are not able to find true compassion, nor create structures of deep transformation, without entering into Jesus' own compassion. Only contemplative-mystical practice, within a context of social action, is capable of bringing about the change of heart necessary for a lasting solidarity – particularly one that embraces the oppressor as well as the oppressed. Thus, according to Galilea, exterior social engagement must be accompanied by a process of interior transformation and liberation from self-seeking. This is the heart of what he terms 'integral liberation'.[47]

However, not all forms of contemplative practice are helpful in this regard. With his broad knowledge of the mystical tradition, particularly sixteenth-century Spanish writings, Galilea has been profoundly critical of a certain kind of Neoplatonic mysticism:

> It has . . . a strongly transcendent orientation and neglects bodily, historical, temporal mediations. It tends to make of contemplation an ascent to God in which the temporal sphere is gradually left behind until an exclusive absorption in God is reached. This tendency can easily become a form of escape.[48]

He calls for a reformulation of the idea of contemplation and of the mystical. At the heart of the tradition, he suggests, has always been the notion of contemplation as a supreme act of self-forgetfulness rather than a preoccupation with personal interiority. In the teachings of the great mystics, contemplation has always been related to the classic Christian themes of cross and death.

> This implies the crucifixion of egoism and the purification of the self as a condition of contemplation. This crucifixion of egoism in forgetfulness of self in the dialectic prayer-commitment will be brought to fulfilment both in the mystical dimension of communication with Jesus in the luminous night of faith, and also in the sacrifice which is assumed by commitment to the liberation of others. The 'death' of mysticism and the 'death' of the militant are the two dimensions of the call to accept the cross, as the condition of being a disciple.

The desert as a political experience liberates [the Christian] from egoism and from the 'system', and is a source of freedom and of an ability to liberate.[49]

The Brazilian theologian Leonardo Boff has sharply criticized the traditional spiritual and monastic formula of *ora et labora* (prayer and work) on the grounds that it espouses a kind of parallelism. At best the *et* has stood for the alternation of interior prayer and exterior practice. Classically, contemplation was the source of all that had value. Practice was not a direct mediation of God but was only of value to the extent that it was 'fed' by contemplation. The whole conceptual framework implies a type of 'spiritual monophysitism': the unique nature of prayer redeems the creaturely and natural profaneness of work.[50] In some contemporary thinking, influenced by a dominance of social and political theory, this parallelism continues to exist but is, as it were, reversed. That is, practice predominates over contemplation so that contemplation becomes another, subsidiary, form of practice. Boff argues for an equal, dialectical relationship 'treating them as two spaces that are open to one another and imply each other'.[51] This dialectic produces a unity in what Boff calls the 'mysticism-politics relationship'. Boff coins a new phrase to describe being contemplative while engaged fully in the public spaces of political transformation – *contemplativus in liberatione*. This unity of prayer-liberation is based on a living faith that 'defines the "from where" and the "towards where" of our existence, which is God and his design of love, that is communicated through, and materialised, in all things'.[52] Thus the contemplative and the mystical 'is not carried out only in the sacred space of prayer, nor in the sacred precinct of the church; purified, sustained and nurtured by living faith, it also finds its place in political and social practice'.[53]

Jürgen Moltmann in his short (and often overlooked) book *Experiences of God* talks of the necessarily ethical dimension of the mystical *sapientia experimentalis* – close to Martin Luther's own understanding of spirituality and to Paul Tillich's related concept of 'participative knowledge'. However, the most interesting aspect of his approach to mysticism is the delineation of a five-fold process to replace the more traditional *triplex via* or 'three-fold way'. This process is really a continuous circular movement. It begins with the human engagement with the ambiguities of the external world. The initial response to injustice, for example, is to act to change things. This action inevitably leads to a realization that a truly Christian response has to be supported by contemplation. Contemplation, which in Moltmann's terms is focused on the history of Jesus Christ in the Gospels, leads to a movement away from self and from false images of God towards 'God alone'. The encounter with the living God, purified of selfishness, is what has been described classically as 'union'. However, this union is not an end in itself. The purpose of mystical union is not to remain in a kind of pure spiritual experience beyond our responsibilities in everyday life. What is encountered at the heart of God is the cross. Union, there- fore, leads to a deeper identification with the person of Jesus who moves out of himself in self-giving love. So, the mystical journey leads the believer back from union, which now becomes a new point of departure, to the practice of everyday discipleship. This is the true meaning of ecstasy, *ekstasis*. It is to step out from oneself in self-giving love:

> As long as we do not think that dying with Christ spiritually is a substitute for dying with him in reality, mysticism does not mean estrangement from action; it is a preparation for public, political discipleship.[54]

The icon of mysticism therefore becomes the political martyr as much as the monk:

> The place of mystical experience is in very truth the cell – the prison cell. The 'witness to the truth of Christ' is despised, scoffed at, persecuted, dishonoured and rejected. In his own fate he experiences the fate of Christ. His fate conforms to Christ's fate. That is what the mystics called *conformitas crucis*, the con- formity of the cross . . . Eckhart's remark that suffering is the shortest way to the birth of God in the soul applies, not to any imagined suffering, but to the very real sufferings endured by 'the witness to the truth'.[55]

What is most destructive in human society arises from the variety of ideologies that claim to contain absolute truth, definitive ethical values or a potent mixture of both. The Christian mystical tradition in the end does point to 'truth', but to a truth that is not concluded, on a way that is no way, in the direction of somewhere that is beyond definitive arrival. The point that the mystical tradition makes is that what is most true can be captured neither intellectually nor affectively.

The journey of perpetual departure

The contemporary Welsh poet R. S. Thomas is fascinated by this inconclusive and elusive quality in the mystical way, the way of unknowing. His poem 'Journeys' expresses the theme of life as perpetual departure:

> The deception of platforms
> where the arrivals and the departures
> coincide. And the smiles
> on the faces of those welcoming
>
> and bidding farewell are
> to conceal the knowledge
> that destinations are the familiarities
> from which the traveller must set out.[56]

Like the writers of so many mystical texts, Julian of Norwich is frequently interpreted in terms of her assurance of the abiding presence and compassion of God in her soul and in the inner depths of her fellow, 'even' Christians. However, Julian is also strongly aware of what I refer to as 'perpetual departure' within the mystical tradition. The Christ whose permanent dwelling is in the heart of the human soul is also the one who always travels ever onwards in pilgrimage. The disciple is not left in comfortable and comforting *stasis*, but is thus drawn after Christ in perpetual dissatisfaction with what is less than everything in pursuit of what Julian refers to as 'bliss in heaven':

Our good Lord revealed himself in his creature in various ways, both in heaven and on earth; but I saw him take no place except in man's soul. He revealed himself on earth in the sweet Incarnation and his blessed Passion, and he showed himself in other ways on earth, where I said that I saw God in an instant of time [a poynte]; and he showed himself in another way on earth, as it were on pilgrimage, that is to say that he is here with us, leading us, and will be until he has brought us all to his bliss in heaven. (LT 81)

Interestingly, there is a contemporary resurgence of interest in mysticism in philosophical and theological circles, in part as a reaction against the notion of absolute metanarratives.[57] The subversive quality of mysticism is represented by the theme of journeys and perpetual departures. This may be thought of as another way in which the mystical dimension offers something in a social, even political world. There is a much greater awareness of the relationship between a post-Enlightenment emphasis on knowledge as objectivity, intelligibility and definition, and issues of power. Thus Michel de Certeau suggested that people whose lives spoke of the 'otherness' of an essentially mysterious God were outsiders to the 'modern' project.

Unbeknownst even to some of its promoters, the creation of mental constructs . . . takes the place of attention to the advent of the Unpredictable. That is why the 'true' mystics are particularly suspicious and critical of what passes for 'presence'. They defend the inaccessibility they confront.[58]

As early as the thirteenth century, that is, since the time when theology became professionalised, spirituals and mystics took up the challenge of the spoken word. In doing so, they were displaced toward the area of 'the fable'. They formed a solidarity with all the tongues that continued speaking, marked in their discourse by the assimilation to the child, the woman, the illiterate, madness, angels, or the body. Everywhere they insinuate an 'extraordinary': they are voices quoted – voices grown more and more separate from the field of meaning that writing had conquered, ever closer to the song or the cry.[59]

De Certeau's interest in mysticism, especially in its sixteenth-century and seventeenth-century forms in the aftermath of the Renaissance and Reformation, concerned a context where the word, especially of Scripture, could no longer be spoken to believers in the old ways. The world was increasingly seen as opaque and unreadable. In response to this disenchantment the people we refer to as mystics sought to invent a different kind of place, that was not a fully formed or constrained place at all. As de Certeau says himself, this 'is only the story of a journey' that is necessarily fragmented and ultimately defies conclusive investigation. In his own somewhat opaque words, 'it overpowers the inquiry with something resembling a laugh':[60] 'That literature offers routes to whomever "asks directions to get lost" and seeks "a way not to come back." Down the paths or ways of which so many mystic texts speak goes the itinerant walker, *Wandersmann*.'[61]

However, mysticism is not a way to replace one ailing system of positivist statement by a new system of intellectual knowledge. It is something other and alien to this form of thinking. It is 'a way of proceeding', a practice, an action. In de Certeau's terms, this terminology of movement implies a continual transgression of fixed points. This whole approach to action was explicitly drawn from his background in Ignatian spirituality:[62]

The various strains of *mystics*, in their reaction to the vanishing of truths, the increasing opaqueness of the authorities and divided or diseased institutions, define not so much a complementary or substitutive knowledge, topography, or entity, but rather a different treatment of the Christian tradition . . . they institute a 'style' that articulates itself into *practices* defining a *modus loquendi* and/or a *modus agendi* . . . What is essential, therefore, is not a body of doctrines (which is the effect of these practices and above all the product of later theological interpretation), but the foundation of a field in which specific procedures will be developed . . .[63]

The later writings of de Certeau concerning mysticism, particularly as it emerged in the sixteenth and seventeenth centuries from the historical ruptures of Renaissance, Reformation and early Enlightenment, appears to make it not only marginalized but also privatized. Indeed it is an act of 'withdrawing' or segregation from social and political place that gives rise to a definable 'mystics'. In de Certeau's own phrase, 'A prophetic faith organised itself into a minority within the secularised state.'[64] Any ambition by the Western Catholic tradition after the Council of Trent to, in de Certeau's words, 'reconstitute a political and spiritual "world" of grace' ultimately failed.

At first sight, this does not appear to offer much room for the social or political nature of mysticism. However, while de Certeau's narrative describes the movement of spirituality to the cultural margins, and its social redistribution in mystic groups with new kinds of discourse and practices, this has to be placed alongside his own postmodern preoccupations and complex thoughts on wandering in reference to Christian faith as a whole.

We should recall that de Certeau himself sought to speak in a world where institutional Christianity was no longer the 'site' of definitive meaning and its status as an exemplary form of human society had been overtaken by the secular state. All that is left, in de Certeau's mind, is the process of following after the perpetually departing Christ. The Christian call is to wander, to journey with no security apart from a story of Christ that is to be enacted rather than objectively stated.[65] De Certeau characterized both the Christian tradition in 'Weakness of Believing' and the withdrawn 'mystic space' in his essay of the same name as 'ways of proceeding' – not institutions but movements, pilgrimages, across defined and definitive places and fixed locations of power. Thus, paradoxically, both Christian practice and mysticism become non-places, disruptive acts of resistance at the heart of all systems and attempts at final statements. Nowadays such resistance is against secular political and economic systems as much as, if not more than, the declining ones of institutional religion.

De Certeau wrote of the 'mystic fable'. It is a fable because it cannot claim the status of definitive truth. It is a language without obvious power. Yet paradoxically, that is its strength. It calls into question the strategically-defined reality of systems of meaning. Believers in Christianity are called in this age to become once again wanderers who are always departing in answer to a call to follow, without the burden of power, authority or even secure identity. The Christian community carries the fabled tale of Christ, especially the empty tomb, which subverts all our places, across an alien territory towards the unnameable that we call 'God'. Following de Certeau we must say that any discourse, not least religious discourse, is always in danger of being shattered. 'Faith speaks prophetically of a Presence who is both immediately felt and yet still to come, who cannot be refused without a betrayal of all language, and yet who cannot be immediately grasped and held in terms of any particular language.'[66]

I do not think it too far-fetched to find echoes here of the later Karl Rahner and his remarks both on the decline of institutional Christianity and the vital importance of the mystical dimension of faith. Thus his famous aphorism not long before he died was that Christians of the future will have to be mystics or nothing at all. We should recall the strong likelihood that de Certeau, like all Jesuits of his generation, had been strongly influenced by Rahner.[67] In his final essays on spirituality and meditations on God, Rahner confronted the 'otherness', the incomprehensibility, of God in a much more explicit way than previously. We find therefore an increasingly radical scepticism, not dissimilar to de Certeau, regarding institutional and intellectual pretensions to certainty. However, as in de Certeau (to my way of thinking), this does not imply that the mystic retreats into protected private places. Discipleship is always bound into the contexts of human history as the loci of self-revelation of the mysterious God. Even as early as the late 1970s, Rahner was prepared to describe the movement of God's Spirit as ranging far more widely throughout the world than the institutional Church would admit. For Rahner, what we term 'grace' is God's implicit inspirit-ing of the world as a whole in ways that drive people onward in a restless movement towards their completion. This is a dispersed mysticism of everyday life in the world of ordinary places. It seems to me that this theme is richer and more complex than a less satisfactory understanding of everyone as 'anonymous Christians':

> If and insofar as the experience of the Spirit I talk of here is also to be found in a mysticism of everyday life outside a verbalised and institutionalised Christianity, and therefore may be discovered by Christians in their lives when they encounter their non-Christian brothers and sisters . . . Christians need not be shocked or astonished at such a revelation. It should serve only to show that their God, the God of Jesus Christ, wants *all* men and women to be saved, and offers God's grace as liberation to *all* human beings, offering it as liberation into incomprehensible mystery. Then the grace of Christ takes effect in a mysterious way beyond the bounds of verbalised and institutionalised Christendom.[68]

Of course, Rahner's language continued to be explicitly Christian in a way that de Certeau's language eventually ceased to be. Yet his continual refusal to build a definitive theological system may be shown to be, in itself, a defence of ultimate mystery and of a kind of weakness at the heart of an essentially mystical view of Christianity. This was expressed more and more strongly as he grew older, not least in these words published posthumously:

> The true system of thought really is the knowledge that humanity is finally directed precisely not toward what it can control in knowledge but toward the absolute mystery as such; that mystery is . . . the blessed goal of knowledge which comes to itself when it is with the incomprehensible one . . . In other words, then, the system is the system of what cannot be systematised.[69]

This apophatic vision, it seems to me, leaves Rahner and de Certeau as, at the very least, distant cousins rather than inhabitants of different intellectual and spiritual universes.

6

Re–Placing the City?

By now it will be clear that my understanding of spirituality focuses less on spiritual *experience,* particularly inwardness, than on *practice.* The first undoubtedly finds its place within spirituality but, I would suggest, only as a manifestation of the second. Equally, I hope that it will be clear that by 'practice' I do not simply mean particular spiritual or devotional *practices* but the practice of living. Of course, for the Christian, at the heart of the 'practice of living' lies discipleship – the intentional following of Jesus Christ that necessitates, as its source, a continual process of conversion or transformation. However, a Christian understanding of spirituality as the practice of living is not confined to the inner life of the Christian community but includes the practice of *everyday* living in the heart of the world of human places. Finally, the Trinitarian foundations of Christian belief demand that 'living' be understood in social or communitarian terms rather than in terms of an essentially individualistic model of personhood.

In this final chapter I want to gather together a number of my thoughts on place and identity in reference to our understanding of the city and the practice of urban living. This is because the city is pre-eminently human in conception and construction. It both represents and creates a climate of values that defines how humans understand themselves and gather together, and also shapes their sensibilities and ways of seeing the world. The future of human cities is an increasingly critical issue. Alongside the problematic question of a rapidly expanding world population is that of the specific growth rate of cities. The figures over the last seventy-five years or so are illuminating. In 1950, 29 per cent of the world's population lived in urban environments. By 1965 this had risen to 36 per cent, by 1990 to 50 per cent. We are reliably informed that best predictions say that this is likely to rise to somewhere between 60 per cent and 75 per cent by 2025.[1] This makes the meaning and nature of cities one of the most pressing human questions that we face. There will never be a single conclusive answer. Indeed, as we shall see, there are those thinkers such as Michel de Certeau who argued fiercely for a politics of resistance to the simple avenues of systematization and homogenization because they see these as destructive forces. However, our always provisional attempts to confront the question of the city will need to involve much more than the principles of economics, engineering, architecture and planning. The city has always been one of the most powerful symbols of how societies understand 'community' and, indeed, the value they place on community at all. The city is where, for an increasing majority of the human race, 'the practice of everyday life' takes place. At base, the question, 'What is a city?' is a philosophical, theological and spiritual one.

I have called this chapter 'Re-Placing the City'. The ambiguity of the phrase reflects a tension both within contemporary philosophical and ecological reflections on the crisis of the modern city and also within Christian theology which has always been dogged by the temptations of a 'replacement eschatology'.

The crisis of place

We have already discussed the contemporary problem of fragmentation and privatization of place and its impact on human identity. Commentators have noted frequently that, during the last fifty years or so, Western societies have undermined place identities in pursuit of values driven largely by economic considerations. In different contexts, the call to economic and social rationalization has produced the contradictory impulses of the centralization of what was once local and the dispersion to multiple sites of what was once a multi-dimensional concentration of living, working and leisure in a single location. The net effect of this has been a growing emphasis on mobility and the growing relativity of space which has dissolved, for many people, the reality of place identity.[2]

In a complex study, Richard Sennett, the eminent American urban historian now at the London School of Economics, blames in part a tendency in Christian theology for the contemporary privatization of space which he sees as a process that began several hundred years ago. Essentially, Sennett argues that modern Western culture suffers from a divide between interiority and exteriority. 'It is a divide between subjective experience and worldly experience, self and city.'[3] This divide, according to Sennett, is based on an unacknowledged fear of exposure. Exposure has the connotation of a threat rather than of the enhancement of life. The result is that, apart from carefully orchestrated spaces for the celebration of heritage or for consumer needs, city building has concentrated on creating safe divisions between different groups of people. Public space thus becomes bland and neutralized as its main purpose is to facilitate movement across it rather than encounters within it.[4] For the city to recover, the need is for the inherent value of the outer life to be rediscovered and reaffirmed.

Sennett blames elements of the Christian theological tradition for this classic inward–outward divide. He interprets Christianity unequivocally as a religion of pilgrimage and dislocation rather than placement. However, the fundamental doctrines of Trinity and Incarnation offer a more dialectical theological perspective on place than he allows. For Sennett, Augustine's *City of God* is the classic expression of the triumph of an inner 'city' in search of eternal fulfilment over the human city.[5] In that Christians continue to have to live in a world of the senses that contrasts with the world of inner, truth-bearing vision, human social places are to be viewed with suspicion. What is most obviously characteristic of these outer places is difference and diversity. Sennett argues that, by denying the outside or, at least, by reducing it to a mere shadow, such a theology underlines the way that Western culture has tended to doubt the spiritual value of diversity. In so far as the valuing of the inner life found expression externally in stone and glass, it was in church buildings or in cathedrals. These, for Sennett, represented a 'place of definition' in cities that actually ran counter to any meaningful definition for the city in itself.[6] Sennett further suggests that modern urbanism stems from what he calls 'a Protestant ethic of space'. This is a further refinement of the Augustinian distinction that he posits – this time in terms of the outer and public as a sterile wilderness. While I agree with much that Sennett says concerning urban culture, his interpretation of Augustine, while not without some justification, is perhaps too sharply drawn. Equally, as we shall see, there are other ways of interpreting the role of cathedrals as the historic centres of cities.

Within the modern city, there has also been a search for purified space, based on another fear that is closely related to that of 'exposure', that is, a fear of mixing and of the disintegration of boundaries.[7] Cities have, therefore, become spaces that exclude. Spatial purification has developed into a key feature of the organization of social space. This is what de Certeau refers to as the creation of 'clean space' in modern urban discourse:

> In this site [the city] organised by 'speculative' and classifying operations, management combines with elimination: on the one hand we have the differentiation and redistribution of the parts and function of the city through inversions, movements, accumulations, etc., and, on the other hand, we have the rejection of whatever is not treatable and that, thus, constitutes the garbage of a functionalist administration (abnormality, sickness, death, etc.).[8]

The theologian Michael Northcott expresses similar sentiments about the disintegration of place in cities. 'The modern city celebrates and facilitates mobility at the expense of settlement, movement at the expense of place.'[9] This is not simply a social issue but a spiritual and theological one. Without a sense of place there is no centring of the human spirit. When human conditions undermine this, the consequent displacement is striking in its effects on individuals and societies. In hardly more than a century, we in the West have moved from a pre-modern, predominantly rural society through an industrial revolution and an urban society into what many people call a post-modern, post-industrial world. In an increasingly placeless culture we have become 'standardised, removable, replaceable, easily transported and transferred from one location to another'.[10] If there is a sense of place, it is predominantly a private one in the face of cynicism about the outer, public world.

We need to be careful, however, in talking in absolute terms about the way that 'mobility' leads to 'anonymity'. All such concepts are culturally and historically conditioned. In the medieval European city, both mobility and anonymity could be viewed as advantages. Migration to the city represented for many peasants freedom from the oppression of feudal ties and anonymity represented liberation from the entrapment of a fixed social landscape and the ability to redefine oneself in more creative ways. In our modern Western cities, the words take on a different value and meaning. They represent the deterritorialization of human life, the compression of space and the homogenization of place. 'The global village creates a depthless and decentred world in which the human identification with locality, place and neighbourhood is often fractured and undermined.'[11]

The privatization of place is also bound up with how we construct our built environments. Architecture and cities are the monuments of our collective consciousness, living symbols of our ideals. When towns began to revive from approximately 1000 onwards, as a result largely of a developing economic role and the growing social importance of the urban mercantile class, two things stand out. First, the urban population continued to need a favourable rural environment because city dwellers were great consumers. Medieval cities were still inextricably part of their surrounding landscapes and the division between urban and rural life was not hard and fast. It is not therefore surprising that in the decoration of medieval cathedrals, for example Chartres, the images of rural life such as the seasons, harvesting and vine-growing still predominate. 'The countryside' was not objectified as a place for leisure. In contrast, today's city is essentially disconnected from the surrounding landscape and sources of food production. Citizens are nowadays global consumers and the supposed limitations of seasonal foods are a thing of the past.

Urban growth in the High Middle Ages also led to the development of the notion that 'the city' could be understood as a holy place. Sometimes this was because of the concentration of religious buildings and artefacts. Italy also preserved the ideal that civic life in itself, with its organized community of people living in concord, could be just as much a way to God as monastic life.[12] The city was often seen as an ideal form of social life that was in effect an image in this world of the ultimate heavenly Jerusalem. There is a whole literary genre, *laudes civitatis*, or poems that articulated a utopian ideal of civic life. Like the glories of the heavenly city, the human city is depicted as a place where many and diverse people are able to live together in peace. This is similar to the classic *pax monastica* that replaced the imperial *pax Romanum*. Then cities were renowned for the quality of communal life in which each and every citizen or group found a particular place that contributed to building up the whole. Finally medieval cities were regularly praised as places of hard work. The point of all this is not merely to show, as medieval scholars now emphasize, that holiness in the Middle Ages was not exclusively bound up with monks or clergy but could be shared by urban communities. The point was also that the city itself was idealized as a utopian vision with a number of key monastic qualities.[13]

By contrast, the monumental architecture that still characterizes much of today's cities stands neither for the value of individual people, nor for intimate relationships, nor again for focused community. Rather, it speaks the language of size, money and power. Commercial complexes such as Canary Wharf tower in London's Docklands exist in brooding isolation rather than in relationship to anywhere else. Our cities frequently lack proper centres that express the whole life of a multifaceted community. The centres of many cities built in the last fifty years, or the centres of old European cities reconstructed since the destruction of the Second World War, have been described as soul-less.

A major part of the problem has been a cellular view of urban planning (originating in part from the philosophy of the French architect Le Corbusier) which has divided cities into 'special areas': living areas, work districts, leisure centres and shopping malls. The immediate consequence has been a fragmentation of the practice of human living and of a sense of diverse community. Such fragmentation is inevitably stressful. On top of this, the creation of the 'city of special areas' has the effect of emptying certain parts of the city at night, especially the centres. This tends to make them dead and even dangerous. Finally, a cellular plan demands the separation of areas from each other by means of distance and clear boundaries. This substantially increases the need for travel and produces more pollution.

In more general terms, this differentiation of discrete areas of activity may be said to reflect a growing secularization of European culture. This has frequently been associated with what is termed the 'commodification' of the city. There is no longer a centred, indeed spiritually centred, meaning for the city, merely a fragmentation into multiple activities, multiple ways of organizing time and space, matched by multiple roles for the inhabitants.[14] Overall, the effect of a cellular form of urban design forces people back into private worlds, or into self-selecting communities, behind defences (physical or spiritual). The designs do not invite people out into shared, humane places of encounter. New domestic ghettos are increasingly protected against sterile public spaces that are no longer respected but, at best, treated unimaginatively and at worst abandoned to violence and vandalism.

Cities reflect and affect the quality of human relationships. The fact is that in the context of urban environments we cannot separate functional, ethical and spiritual questions. If place is to be sacred, places must affirm the sacredness of people, community and the human capacity for transcendence. I would argue, against Sennett, that in an earlier age the cathedral fulfilled that function in Western society. It was at the same time an image of God and a symbol of the ideals of the citizens at the heart of the city.

In a sense, the cathedral offered a focus not simply for a two-dimensional pattern of the city – its static 'map' or 'grid. There was a third dimension, that of movement through space that was not generated merely on the horizontal plane but also upward. Indeed, the cathedral even spoke of a fourth dimension – that of time and especially of transformation through time. In his attempt to describe an urban aesthetic, the philosopher Berleant suggests that the role of the cathedral was as a guide to an 'urban ecology' that contrasts with the monotony of the modern city 'thus helping transform it from a place where one's humanity is constantly threatened into a place where it is continually achieved and enlarged'.[15] Such an urban 'centre' offers communion with something that lies much deeper than simply the need for regularity and order in shared public life. It is not purely functional but evocative. If we leave behind for the moment the kind of specialist theological language that we must inevitably use of such an explicitly *Christian* religious symbol, it is possible to speak in more general terms of what something like a cathedral achieves in its role as a heart for the city. So, for example, it deliberately speaks of 'the condition of the world'. It both expresses the history of human experience and yet transcends easy understanding. Perhaps most important of all, it is a repository for the memory and the aspirations of the community which have been constantly renewed and changed across time. To enter such a building is to enter into communion with centuries of human pains, achievements and ideals. Indeed, the moment a building like a cathedral becomes *fixed* rather than something fluid and continually changing it is a museum rather than a living symbol of the city. If the cathedral presented, in architectural form, a living symbol of the ideals of a community, what has replaced it now that the dominant social institutions have shifted from a religious sphere to the secular?

Too often, in fact, our contemporary urban places no longer have this centred quality because we have built nothing into them that is truly precious to us. In the past material places had a clear and recognizable order. Buildings were so organized that they were both significantly differentiated from each other and, at the same time, sufficiently related to each other. Place is space that has the capacity to be remembered and to evoke our attention and care. We need this if life is to be conducted well. 'We need to think about where we are and what is unique and special about our surroundings so that we can better understand ourselves and how we relate to others'.[16]

It is not unusual to regard the modern city as a purely functional environment. Yet even 'function' involves more than simply practical organization. The issue of space is more than an impersonal problem of engineering. The issue of urban space has a great deal to do with the creation of perceptions. The height of the skyscraper is not the same kind of elevating moral and spiritual presence as the cathedral. It tends to speak of the forces of sheer size, of economics and power that are perceived as oppressive and impersonal. If we think of the cathedral less in a narrow literal sense but more as a paradigm, such a building suggested that, at its best, our sense of place is a sense of the sacred – sacred to people and sacred in relation to a higher order, however that is conceived. The fact that we may find it incongruous to think of our built environments as having a sacred quality merely suggests the degree to which the modern city has signally failed to become precious to its citizens. Indeed, has precisely failed at heart to create a sense of *citizenship* as opposed to a mere sense of residence.

The eminent French anthropologist, Marc Augé (a pupil of de Certeau) writes about the differences between traditional French towns and the new towns 'produced by technicist and voluntarist urbanisation projects'.[17] Traditional towns have not only aspired to be the centre of somewhere (a region) or something (of gastronomy, for example) but also have since the Middle Ages developed a monumental centre to symbolize and materialize their aspirations. The smallest of towns and villages boast a 'Place' or town centre. These contain the buildings that symbolize authority and meaning, whether religious (the cathedral or parish church), or civil (the *hotel de ville* or *mairie*) and sometimes a historical monument (for example, the war memorial). These buildings tend to overlook an open space through which many of the cross-town routes pass. The key feature, however, is that such town centres were and are *active* places. People gather there. The leading cafés, hotels or businesses concentrate as close to the square as possible. The problem with the new towns, according to Augé, is that they fail to offer 'places for living'. By 'places for living' he means contexts, 'where individual itineraries can intersect and mingle, where a few words are exchanged and solitudes momentarily forgotten, on the church steps, in front of the town hall, at the café counter or in the baker's doorway'.[18]

Responses to the crisis of place

There have been a number of different approaches in response to the crisis of place. One of the religious philosophers of the American political Right, Michael Novak, a once liberal Roman Catholic, finds a strange and disturbing theological justification for the kind of uncentred culture that gives birth to the sterility of centreless cities. Essentially, the virtue of pluralism that Novak believes should characterize Western 'democratic capitalism' demands that each of us be our own centre in a world of private preference – if we actually have that luxury! Because conscience must be free of socially produced constraints, there can be no agreed spiritual core or imposed notion of 'God'. For Novak, what he refers to as an 'empty shrine' at the heart of society stands, to my mind perversely, for a proper reverence of the transcendent. For Novak, theologically speaking, the central Christian doctrine of Incarnation implies that we must respect the world as it is rather than preach some kind of process of social transformation of the *outer* world. In fact this is a theology of the Incarnation evacuated of its redemptive heart. Such a society, empty of shared conviction and prey to those with economic, social or political power, merely produces a series of urban deserts from which 'all shared narratives of human hope' are absent.[19]

The writings of the British theologian John Milbank are a long way from the nihilism of Michael Novak. His project is to salvage a positive theological vision from sterile secularity and from what he believes to be an unhelpful subordination of theology to the new orthodoxy of the social sciences. Unfortunately, in doing so, Milbank's position appears to me to run a risk of perpetuating rather than healing the division between an outer (material) and inner (spiritual) city.[20] Milbank writes of the Christian vision of a city or a society of peace rather than of violence. Milbank would, I suspect, have no sympathy for Novak's defence of a contemporary 'wasteland' in which the fit survive and the remainder are casualties. However, in the end Milbank's vision appears to be equally, if differently, bleak.

The question is whether this vision is too protected and too otherworldly to be redemptive in the world of places and events. The outer city is conceived of as essentially 'secular', that is, irredeemably immersed in the *saeculum*, the here and now, and is to be rejected because, built on human reason, it is inherently involved in an ontology of violence. I am not convinced that, in the end, this is pure Augustine at all. The view comes closer to the austere ethical Protestantism of someone like the French thinker Jacques Ellul than it does to a critical and prophetic sacramental sensibility of a Rowan Williams – or, indeed, of a Duns Scotus. I am concerned that such a vision may not be capable of *redeeming* time and place, of expressing the sacramental nature of the world of particular places, and of transforming the so-called 'secular city'. The life of the Christian community seems too sharply set apart from the contingent world.

While I agree that the Church is a political reality,[21] there is a danger in relocating true politics entirely to the inner life of the Church. For one thing, the Church is never a perfect politics. I also agree with Milbank's view that ecclesiology is really concerned with a structural logic for human society.[22] However, the precise role of the Church *within* society is not clearly addressed. This is difficult because it seems to me that an important strand within a Catholic ecclesiology itself critiques any temptation to ecclesiolatry by suggesting that the Church is only truly itself when it is broken open on behalf of the world in the midst of the world. It may be a prophetic, disruptive or heterological space on the margins of the 'world city', but it misses its vocation if it seeks to make itself a self-contained ethical space fully set apart. Such a space would not be Augustine's Other City. There appears to be a dangerous lack of clarity in the end about the importance of a distinction between ecclesial theory and the contingency of the institutional Church. Indeed, contingency of any sort seems to be the prerogative solely of the secular, pagan and fundamentally sinful city. The danger is that this vision 'effectively destroys the idea of a city'.[23]

Because Augustine's *City of God* was more concerned with the city as community (*civitas*) rather than physical place (*urbs*), people have been able to draw from it a radical distinction between the earthly and heavenly cities. This distinction was reinforced in the

Etymologiarum libri by the seventh-century Isidore of Seville. He described *urbs* and *civitas*, 'cities of stone; cities of men', as existing on two separate planes without necessary interaction. This image remained etched in the minds of people until the late Middle Ages and, arguably, into the modern era:

> A city [*civitas*] is a number of men joined by a social bond. It takes its name from the citizens [*cives*] who dwell in it. As an *urbs* it is only a walled structure, but inhabitants, not building stones, are referred to as a *city*.[24]

In practice, of course, there needs to be a dialectical relationship between the two planes. Only then will there develop a community-centred plan for cities that gives reason for the physical space by expressing the ways that life is actually lived, or that people hope that it may be lived, rather than by reference only to a choreography determined purely by 'planners' or urban theorists.

Essentially, the true 'city' for Augustine was the community of believers that was destined to become the City of God. Within the human city, this community could be seen as set apart from the stream of the everyday (the Church). Or, more properly speaking, it was hidden entirely given that the human Church contained those who might not make it into the Kingdom and that many people would, by God's grace, make it into the Kingdom without benefit of clergy, as it were.[25] Clearly Augustine was rightly suspicious of any attempt by Christian rulers to suggest that their Christian commonwealth was somehow the Kingdom of God proclaimed by Jesus Christ or at least its first cousin. This aspect of Augustine's legacy gives Christianity a great deal of prophetic ammunition to attack any attempt to divinize economic and political systems (for example, capitalism or Marxist-Leninism) or particular empires (for example, the thousand year Reich). However, it leaves Christianity commensurately weak in offering positive visions for reconstructing human societies in Christian terms or for reclaiming public place as sacred. This is a legacy that Christianity still struggles to overcome.[26] However, to be fair, the individualism and emphasis on privacy and interiority that permeates so much of

Western culture would have been alien to Augustine. In his literal commentary on the book of Genesis, the root of all evil was self-enclosure or privacy. The 'private' was the opposite of 'shared', 'common', 'public'. The Heavenly City was to be a community in which the fullness of sharing would be had. For Augustine it is humanity, rather than autonomous individuals, that is created in the image of God. Virtue consists of defending what is public or held in common. There will be no room in the Kingdom of God for a self-enclosed and protected privacy.[27]

As I have already suggested, Augustine's theory of history belongs in the eschatological category. In that sense, the City of God operates in a realm distinct from everyday history. However, the distancing is far from absolute. True, no human *polis* is the Kingdom of God in human guise. The Earthly City is undoubtedly contingent. But human history is God's creation and is not, therefore, to be condemned as merely evil. Augustine possessed a deep sense that the world of places and of *each moment* was equally filled with God's active presence. Augustine's theology of history essentially describes a thread of godly history running through the history of contingent events in the world of particular places.[28] Augustine's distinction between sacred and secular cities does not render the history of human, contingent places meaningless. What Augustine rejects is any sense that the contingent world or human politics is of ultimate value.

Resistance

We tend to think that harmony, a harmonious arrangement of our human environment, simply implies *order*. Yet, part of the aesthetics of a healthy city, which contrasts with a purely efficient mechanics, is the way it facilitates the transcendence of absolute order:

While there is formal structure in a quartz crystal and a starfish, as there is in the symmetry of the Taj Mahal and Notre Dame Cathedral, art, like nature, has its share of deliberate disarray. We can find as much disorder in the opening movements of

Bach's great organ toccatas in C major and D minor and in Debussy's through-composed songs as in the irregular curve of a beach or the scattering of daisies in a field.[29]

The kinds of space theory that planners can impose on city environments in order to 'make sense' of them are frequently totalitarian. In writing about 'Walking in the City',[30] de Certeau found another way to express one of his favourite themes, that of 'resistance' to ordered systems that leave no room for otherness and transgression. The weak, in this case, those who actually live in the city, find ways to make space for themselves and to express their self-determination. What de Certeau calls 'the urbanistic system' defines a 'literal meaning' of geometrical space that is similar to those 'proper meanings' in the use of language constructed by grammarians! Against this, there are the people who actually walk in the city. This dimension is what he calls the 'noise', in other words the 'difference', the 'otherness' that is a city's life blood and without which it will die or become an empty shell. That is why, in what I would call de Certeau's integral aesthetic of a city, the role of indeterminacy is so important. He refers to this in terms of 'casual time'. 'Thus to eliminate the unforeseen or expel it from calculations as an illegitimate accident and an obstacle to rationality is to interdict the possibility of a living and "mythical" practice of the city.'[31]

Like Paul Ricoeur, de Certeau placed a strong emphasis on the power of narrative to shape human environments and to transform them. Indeed, in terms of everyday life in the city, it is narrative as much as architecture or the planning of environment that shapes identity and enables people to *use* the city as a means of creative or effective living:

> In modern Athens, the vehicles of mass transportation are called *metaphorai*. To go to work or come home, one takes a 'metaphor' – a bus or a train. Stories could also take this noble name: every day, they traverse and organise places; they select and link them together; they make sentences and itineraries out of them. They are spatial trajectories.[32]

De Certeau also stressed the importance of narrative to the practical articulation of everyday actions. Stories are more than descriptions: they also take ownership of spaces and are therefore culturally and socially creative. Because human stories define boundaries, and also create bridges between individuals, narrative is also a vital factor in the creation of the city as a community rather than merely as an agglomeration of buildings and spaces. The narrative structure of such communities enables people to shape the world that surrounds them, rather than be passively controlled by it, and also creates ways of mapping the city and thus moving around it effectively.[33]

The subversive power of children

An important factor in the resistance to the homogenization of city culture is the space children occupy or are allowed to occupy. 'Today, talking about the space of children is, most of all, talking about its absence.' Thus commented sadly Professor Franco LaCecia in his 1999 lecture 'The Space of Play' at the Royal Society of Arts. The place of children is one of the most potent indicators of how, in reality, the city is conceived and practised.[34] As a number of commentators have noted, the history of the modern city over the last fifty years has involved the progressive abandonment of most of the spaces that were once available to children for bodily movement. Arguably one of the largest factors has been that of the problems of safety. However, the concept of 'safety' itself is largely the product of the move increasingly towards the development of monofunctional spaces where every activity is assigned its place and every space is defined in terms of specific function. This means that city spaces are frequently not 'possessed' or walked in human bodily terms at certain times (outside work hours), by certain categories of people (the 'vulnerable'), or at all.

In contrast, children have always represented an apparently chaotic and unpredictable way of using urban space. But is this not merely 'play'? In Western, 'grown up' cities, 'play', if it has any value at all, is increasingly one activity among many that needs to be confined to its own monofunctional space, called, for example, a leisure centre. Such centres, interestingly, are usually situated away from the centre on the periphery of town. In other words, 'play' is marginalized from the heart of cities as something that does not fit in terms of the dominant values of efficiency and safety. 'Play' has no discernible productivity and does not easily find a place within a results-oriented culture. It seems to be on those terms that some New York City schools have recently abolished play time! The status and role of children, therefore, particularly their invisibility except as instruments in promoting consumer concepts (to quote one telling journalistic phrase, 'from person of the future to handbag of now'), is not merely a statement about cities but also about our concept of human personhood. This is reduced to rationality, efficiency and productivity. Certain kinds of non-productive people (including children) become a 'liability' and therefore a worry.

Perhaps to reduce children to the stereotype of 'those who play' is itself a problem. 'Play' seems to be an adult category that neatly bypasses how seriously and with what complexity children experience and practise their inner and outer worlds. What adults describe as 'play' is, for children, 'a practice of everyday life'. In reality, children are amongst the most powerful challenges to the city as *planned space*. They automatically transgress boundaries and will insist in promoting a multi-functional approach in contrast to the adult preference for monofunctional spaces. Adults think about the violence and the traffic and children's safety (along with that of other so-called vulnerable or non-productive groups) and yet miss the underlying question, 'What is a safe city?' In the urban thinking of people such as de Certeau and his pupil Augé, the truly human city is a place of embodiment rather than a place defined by abstract concepts. In so many ways, children stand for an embodied rather than purely mental practice of everyday life. In other words, the truly human city is where the human body, and the collective body of the citizenry, in all its modes dominates; where the streets are still *walked* in ways that lack normally defined functionality and non-planned space invites the creativity present in a multitude of human possibilities:

Cities are safe only as long as streets are a place for human living: crowded, multifunctional, open to every kind of passing and staying. Only an inhabited street can avoid here the danger of an inhuman city.[35]

Recovering the city

We could turn our backs on the city in pursuit of a rural idyll. However, apart from the logistical impossibility for the majority of such a massive social reversal, the danger is that unless we solve the problem of alienation at some other level we simply carry it with us to another place. We do not need so much to flee the city as to repossess it for people, *all* people, by day and by night. The state of our urban environment highlights again the importance of ethics to any reflection on place. Whose place is it? Who owns it? Who is kept out or marginalized? Who is not made to feel at home?

In fact what we need to confront the fundamental question, 'why cities – what are cities for?'[36] In our Western world they no longer have a strictly practical role as a defence against attack or a refuge against wild beasts. Even the more recent rationale for cities during the last two hundred years – the city as focus for economic systems and the organization of production – are ceasing to be as necessary as they were. The decentralization of industry and the as yet only half-sensed impact of internet living mean that the complexity of cities is less and less necessary for efficient work or for the distribution of goods. Information technology is also likely to confront the previously unquestioned role of the city as *the* centre for information and for education. I do not believe that this means the end of the city but the beginning of a new phase.

In the future, the city is most likely to find its meaning within the wider requirements of human culture. There will have to be far greater reflection on the civilizing possibilities of the city and the opportunity it may offer for community, or social, humanization and cultural meaning. This implies that the aesthetic and moral potential of cities will take over from economic or other functionalist purposes. Such potential includes a number of things that, in terms of immediacy, are difficult to pin down and are yet critical.

Cities have a unique capacity to focus a range of physical, intellectual and creative energies. They create new sets of cultural relationships simply because cities bring into regular contact a uniquely diverse range of activities. Cities have an unparalleled ability to combine diversities of age, ethnicity and cultures. Because of their relatively large size *and* the diversity of their spaces, cities are able to balance community and anonymity. As human environments, cities when they work effectively have an ability perpetually to undermine the defensive human instinct for enclosed assumptions and foregone conclusions.[37]

This points increasingly to a new, or rather revived and deepened central question: What is a humane or humanizing city? Theologically speaking, if the human city is merely an unavoidable environment for transit to somewhere else, the eternal city of God, it does not need an aesthetics but merely a pragmatics or a mechanics. A certain interpretation of Augustine's theology of history leaves us with the sense that the city does not describe anything essential about what it is to be human but is no more than an ephemeral shell for nomadic pilgrimage. By contrast, in the humanized city people would not only dwell but also belong – that is to say, be joined in attachments of affection and fulfilment. The humane city would enable the necessary space for individual personality to be balanced with a healthy collectivity. It would enable human aspirations to be productive rather than either repressed or diminished into self-indulgence. It would enable a proper connection to the world of nature such that 'environment' or 'ecology' would not be distanced or placed in opposition but be continuous with and integrated with the human person.

The virtue of postmodern approaches to cities by architects, city planners and anthropologists is that these seek to rediscover place, to affirm human difference, to facilitate the recovery of territorial identity, to build community, to reshape public space and to plan built environments in accordance with human proportion rather than in terms of standardization and depersonalizing bulk. However, we need to be careful. Such postmodern approaches also fail to engage with what one might call the structural sins of power dominance or excessive economic differentiation. Without an integral approach to the recreation of human community, it is very difficult to prevent the creation of new public spaces, cleaned up city centres and the pavement café culture from once again insulating affluence from a deprivation (social and cultural as much as purely economic) that has not been fundamentally confronted. The result is all too easily to reconfirm new forms of exclusion.

What is demanded, if cities are to have a long-term and meaningful future, is a reversal of the trend for them to represent merely a necessary evil. Traditional justifications (work, commerce, information and education) are likely to become increasingly *unnecessary* or avoidable for those who can afford it. A long-term future demands the replacement of perceptions such as alienation, isolation, crime, congestion and pollution by community, participation, human energy, aesthetics and joy. In his much commended Reith Lectures of 1995, the architect Richard Rogers summarized the current situation:

> The city has been viewed as an arena for consumerism. Political and commercial expediency has shifted the emphasis of urban development from meeting the broad social needs of the community to meeting the circumscribed needs of individuals. The pursuit of this narrow objective has sapped the city of its vitality. The complexity of 'community' has been untangled and public life has been dissected into individual components. Paradoxically, in this global age of rising democracy, cities are increasingly polarising society into segregated communities.[38]

Rogers pleaded for the recovery of the concept of 'open-minded' space that he borrowed from political theory. This is multi-functional and makes moral and spiritual, as well as literal, space for a variety of uses by a variety of people in which everyone becomes a participant. This kind of space contrasts with the current dominance of 'single-minded' space that has one function predetermined from outside the environment and its participants by designers and planners. Both spaces may be necessary in any given urban community. 'Single-minded' space responds to a modern need for autonomy and efficiency. However, without 'open-minded' space as well there are no places that give the inhabitants something in common or bring together a plurality of people.

Theological reflection on cities in recent years has tended in England to focus largely on what have been called, particularly in Anglican terms, 'urban priority areas'. This may actually produce an unbalanced result. If there are sinful structures of exclusion and social deprivation these are not limited to particular districts within cities but effect, perhaps I should say 'infect', the city as a whole both as built space and human community. If there is a message of liberation and transformation that the Christian gospel proclaims, it must be an integral one for the concept of the city as a whole. 'Urban priority areas' are what they are, socially and economically, because of wider cultural failures concerning the nature of what it is to live publicly and the definition of human life as interdependence, the lack of a philosophy of humane environments, of community and the like. As a corporate expression of human self-definition, the city as a whole is a statement about the boundaries and potential of what it is to be human. What does it mean to be humane as well as human? What does human community amount to? Are equality and personal freedom fundamental values? If so, why and how far can they be achieved?

There is a growing sense that public space needs to be reclaimed for all the citizens. Historically urban culture has 'worked' only when it has been fundamentally participatory. This sense of democratic participation demands physical expression and for this an 'open-minded' public realm is indispensable. 'Open-minded' is not merely about size for it is noticeable that the great totalitarian regimes of the twentieth century on Left and Right adopted the sheer vastness, and monumental quality, of public space to intimidate and to control the individual. Public space needs to be accessible not only physically but also 'intellectually' in the sense that, in its design, it should address the questions of inclusivity for all citizens and of safety at night as well as by day.

In his definition of the sustainable city of the future, it is interesting that Richard Rogers adopts seven principles that are spiritual as much as or more than purely functional. A city will need to be just (fundamentally accessible to all and participative), beautiful (with an aesthetic that uplifts the spirit), creative (able to stimulate the full potential of all its citizens and able to respond easily to change), ecological (where landscape and human action are integrated rather than in competition), 'of easy contact' (where communication in all senses is facilitated and where public spaces are communitarian), polycentric (integrating neighbourhoods and maximising proximity), and finally diverse.[39]

Conclusion: theological reflections

There are serious theological questions which we must reflect upon if our built environments are to support the kind of relationships that will enable humans, individually and collectively, to achieve their deepest identity. Culturally speaking, places are constructed by the people that inhabit, use or relate to them. There is another biblical image of the city apart from the one in Genesis of Cain the murderer becoming the first city dweller or the people of Babel seeking to replace the authority of God. This lies in the Jerusalem tradition, for example in the Psalms. Here the city is to express the peace of God. Those who live in the city are required to share God's peace with one another:

> Hence the city is called to realise a justice which is more than giving each person his due: it is to give God his due, by building the city that his peace, presence and forgiveness make possible, so that all his people may share it together.[40]

From a Christian standpoint, people are created in the image of a *Trinitarian* God. Trinitarian faith has a particular capacity to hold unity and multiplicity together. This offers a way beyond the contemporary tendency (reflected in much postmodern writing) for community and indeed the individual 'self' to fragment into a 'pastiche of personalities'. The Christian revelation of God suggests a dialectical relationship between particularity and universality, or between what is personal and what is interdependent. Christian Trinitarian belief affirms God as a space in which the particularity of the divine persons is not merely held in tension with, but consists of their interrelatedness and communion. The portrayal of God as persons-in-communion can be translated into an image of human well-being as 'personal space' that is indistinguishably 'space for the other'. There are political implications with regard to a vision of citizenship in the city. A Christian vision necessarily affirms both a variety of life and also the common good (1 Cor. 12). The true depth of human good consists of a particularity whose very existence depends on openness to what is other, beyond and more.

The Christian story unequivocally relates that we come to be as people in and through communities of reciprocal and equal relationships. These relationships both reflect and deepen our contact with the 'community' that is God. In terms of this theological vision, our built environments can be either sacraments or anti-sacraments. They may be revelations of God or denials of God. As such they also reveal what it is to be human or become false 'revelations' of the underlying meaninglessness of human existence. If what we build is an antithesis of human proportion, we should not be surprised that the lack of either intimacy or glory radically undermines the image of the divine at the heart of human living.

There is a need to ask what it means to be truly human and thus what human well-being is. There is also a need to help society define community not simply in terms of a satisfying glow but in terms of the hard road of reconciliation. We have already noted that reconciliation is more than mutual accommodation. It is a costly matter, because it goes much further and deeper. There is that *Oxford Dictionary* definition of reconciliation as 'the reconsecration of desecrated places'. This refers both to the place of the individual person who has been marginalized or diminished and the locations within our cities, often the centre-less centres as much as run down estates, that speak only of a human void. A place of reconciliation does not homogenize people or environments but creates space for the diversity of human voices to participate. Most of all, as we have seen, a space of reconciliation invites all who inhabit it to make space for 'the other', to move over socially and spiritually, to make room for those who are unlike, and in that process for everyone to be transformed into something new.

The Eucharist is, in Christian terms, the most potent symbol of making reconciled space, catholic space and ethical space. I want to offer an image from the urban past. In the medieval city, the Eucharist was the centre of life in the midst of the city not merely as a form of private devotion but as a *public drama* that enacted Christ's presence that was daily incarnated in the heart of the city. Thus when processions of the Blessed Sacrament took place through its streets on the feast of Corpus Christi (whatever people may nowadays think of the underlying theology behind the ritual), a city like London 'is not only a physical community but also a host of angels singing "Holy, holy, holy!"'[41] The Eucharist becomes the *genius loci* – the true meaning of the place. Similarly the performance of Passion plays in the streets, the presence of relics, shrines and holy wells even in cities like London spoke of a sense of place and time that is at odds with the realities of a twenty-first-century city as an impersonal, smoothly running mechanism. A sense of the sacredness of place was strongly present. This was not limited to the oases of church buildings but included the ordinary spaces and streets. On Rogation Days there was a ceremony of beating the bounds. In other words, the boundaries of each parish were marked out by a religious procession. This symbolized the freeing of a sacred space from the spirit of evil. This did not mean that there was no sense of the sinfulness of the streets. However, medieval society was much more at home with a dialectical vision of reality that could encompass the sacred and the sinful at the same time and in the same space. Our greater sense of the separation between light and dark is in many ways a reflection of a post-Enlightenment world-view.

There is no point in pretending that we can recreate, in our present Western society, the Eucharist as a public drama *in this sense*. To pretend otherwise would be a futile exercise in nostalgia. Nevertheless, the essence of the Eucharist remains a public rather than private drama. There is another side to the theology and practice of the Eucharist that moves us beyond immediate issues of the public presentation of the ritual and we often overlook this. Does eucharistic theology only allow the Eucharist to be 'Church space' or do we have a theology that permits the Eucharist to be,

more broadly, a space for the world? What is really being enacted
when the Eucharist is celebrated in the midst of the human city?
Church people are frequently concerned by the need to make the
liturgical celebration more accessible to the community it serves,
more participative, more expressive of the actual values, experiences and stories of the worshipping community. For some
eucharistic theologies, particularly those that emphasize the
Eucharist as 'enacted Word', there is a deep question concerning
the effectiveness of this word. Is it a word that must be actively
heard by believers and then consciously responded to? Or does the
word of God reach out beyond those capable of receiving it in this
immediate and obvious way to accomplish something far more than
we can understand now?

All Christians would affirm, I presume, that every celebration of
the Eucharist by a particular community is, in its catholicity, more
than a local possession. I would go further and suggest that the
Eucharist is also always the possession of more than the visible
Church. Of course, a Eucharist celebrated in a virtually empty city
church raises painful pastoral questions about the mission of the
institutional Church. Equally, it is easy to slide unconsciously into
magical ways of thinking about the Eucharist – that it produces
its wonders simply by the enactment of ritual irrespective of
attendance figures. Yet the question of a universal meaning for the
Eucharist remains. There is a strange yet powerful dynamism, it
seems to me, in the solitary eucharistic celebrations of someone like
Charles de Foucauld during his years as a hermit amongst the
Tuareg of Morocco just as there is in the Eucharists of the small
fraternities who follow his way of contemplative presence in
today's inner cities.

Again, was Pierre Teilhard de Chardin's concept of the cosmic
nature of the Eucharist, immortalized in his extraordinary 'Mass on
the World', merely some form of rather self-indulgent pantheistic
experience? On the contrary, I believe it touches the heart of the
matter. As is evident from other texts, Teilhard was unusual in an
age when popular Roman Catholic eucharistic theology was still
strongly individualistic. He not only had a developed theology of
the mystical body of the Christian community but also a theology

of the 'prolongations' (his own concept) or extensions of the Eucharist into the whole inhabited world and into the mystery of the cosmos. 'From the particular cosmic element [the eucharistic elements] into which he [Christ] has entered, the activity of the Word goes forth to subdue and to draw into himself all the rest.'[42] Teilhard was deeply preoccupied by the planetary dimensions of his normal daily celebration of the Eucharist. These same thoughts are expressed beautifully in the Offertory Prayer of 'Mass on the World' inspired by his experience on the steppes of Asia at dawn of the Feast of the Transfiguration (his favourite feast), 1923. Here it seems good to end:

One by one, Lord, I see and I love all those whom you have given me to sustain and charm my life. One by one also I number all those who make up that other beloved family which has gradually surrounded me, its unity fashioned out of the most disparate elements, with affinities of the heart, of scientific research, and of thought. And again one by one – more vaguely it is true, yet all-inclusively – I call before me the whole vast anonymous army of living humanity; those who surround me and support me though I do not know them . . . This restless multitude, confused or orderly, the immensity of which terrifies us; this ocean of humanity whose slow, monotonous wave-flows trouble the hearts even of those whose faith is most firm: it is to this deep that I desire all the fibres of my being should respond. All the things in the world to which this day will bring increase; all those that will diminish; all those too that will die: all of them, Lord, I try to gather into my arms, so as to hold them out to you in offering. This is the material of my sacrifice; the only material you desire . . . Receive, O Lord, this all-embracing host which your whole creation, moved by your magnetism, offers you at this dawn of a new day. This bread, our toil, is of itself, I know, but an immense fragmentation; this wine, our pain, is not more, I know, than a draught that dissolves. Yet in the very depths of this formless mass you have implanted . . . a desire, irresistible, hallowing, which makes us cry out, believer and unbeliever alike: 'Lord, make us one'.[43]

Notes

1. A Sense of Place

1. Donlyn Lyndon and Charles W. Moore, *Chambers for a Memory Palace*, Cambridge, MA: MIT Press 1994, p. xii.

2. Philippa Berry, 'Introduction', pp. 1–2, in Philippa Berry and Andrew Wernick (eds), *Shadow of Spirit: Postmodernism and Religion*, London: Routledge 1992.

3. Michel de Certeau, *The Mystic Fable*, ET Chicago: University of Chicago Press 1992, p. 299.

4. Clifford Geertz, *The Interpretation of Cultures*, New York: Basic Books 1973, pp. 4–5.

5. Geertz, *The Interpretation of Cultures*, p. 89. Geertz' emphasis on the symbolic nature of culture has commended his approach to writers who seek to reflect theologically on culture. See, for example, Michael P. Gallagher, *Clashing Symbols: An Introduction to Faith and Culture*, London: Darton, Longman & Todd 1999, and Kieran Flanagan, *The Enchantment of Sociology: A Study of Theology and Culture*, London: Macmillan/New York: St. Martin's Press 1999.

6. Geertz, *The Interpretation of Cultures*, p. 14.

7. A. J. Gurevich, *Categories of Medieval Culture*, ET London: Routledge & Kegan Paul 1985, pp. 4–5.

8. A good summary of these issues can be found in the editors' introductory essay 'Culture, Power, Place: Ethnography at the End of an Era', in Akhil Gupta and James Ferguson (eds), *Culture, Power, Place: Explorations in Critical Anthropology*, Durham, NC: Duke University Press 1999, pp. 2–29.

9. Gurevich, *Categories of Medieval Culture*, p. 94.

10. See George Lakoff and Mark Johnson, *Metaphors We Live by*, Chicago: University of Chicago Press 1981, particularly Chapter 4, 'Orientational Metaphors'.

11. Christopher Tilley, *Metaphor and Material Culture*, Oxford: Blackwell 1999, p. 177.

12. See Mircea Eliade, 'Sacred Places: Temple, Palace, "Centre of the World"', in *Patterns in Comparative Religion*, New York: World Publishing 1963; Victor and Edith Turner, *Image and Pilgrimage in Christian Culture*, New York: Columbia University Press 1978; a critique in John Eade and Michael Sallnow (eds), *Contesting the Sacred: The Anthropology of Christian Pilgrimage*, London: Routledge 1990, Introduction, *passim*. On conflict, see Simon Coleman and John Elsner, *Pilgrimage: Past and Present in the World Religions*, London: British Museum Press 1995, Picture Section II 'The Sacred Site: Contestation and Co-operation', pp. 48–51.

13. On these issues see 'Introduction', in Steven Feld and Keith H. Basso (eds), *Senses of Place*, Santa Fe: School of American Research Press 1996, pp. 3–11.

14. See Edward S. Casey, 'How to Get from Space to Place in a Fairly Short Stretch of Time: Phenomenological Prolegomena', in Feld and Basso (eds), *Senses of Place*, pp. 13–52. Also Gaston Bachelard, *The Poetics of Space*, Boston: Beacon Press 1994.

15. Martin Heidegger, *Poetry, Language, Thought*, ET New York: Harper & Row 1975, p. 154.

16. 'An Ontological Consideration of Place', p. 26, in Martin Heidegger, *The Question of Being*, ET New York: Twayne Publishers 1958.

17. Heidegger, *Poetry, Language, Thought*, p. 157.

18. Walter Brueggemann, *The Land: Place as Gift, Promise and Challenge in Biblical Faith*, Philadelphia: Fortress Press 1977; London: SPCK 1978, p. 5.

19. Arnold Berleant, *The Aesthetics of Environment*, Philadelphia: Temple University Press 1992, p. 4. On the humanly constructed meaning of place see also Simon Schama, *Landscape and Memory*, London: HarperCollins 1995, for example pp. 6–7, 61, 81. Schama illustrates his point particularly powerfully in Part 1 'Wood', where he discusses the different cultural and even spiritual values associated with the forest in Europe and North America.

20. See Anne Buttimer, 'Home, Reach and the Sense of Place', in Anne Buttimer and David Seamon (eds), *The Human Experience of Space and Place*, London: Croom Helm 1980, p. 174.

21. Berleant, *The Aesthetics of Environment*, pp. 86–7.

22. Marc Augé, *Non-Places: Introduction to an Anthropology of Supermodernity*, ET London/New York: Verso 1997, especially pp. 51–2, 77.

23. On deterritorialization, see Arjun Appadurai, *Modernity at Large: Cultural Dimensions of Globalisation*, Minneapolis: University of Minnesota Press, 1998, *passim* but especially Chapter 9, 'The Production of Locality'.

24. Appadurai, *Modernity at Large*, pp. 29, 178.

25. Brueggemann, *The Land*, p. 4.

26. Simone Weil, *The Need for Roots*, ET London/New York: Routledge 1997, p. 41.

27. See Heidegger, 'Building Dwelling, Thinking', in *Poetry, Language, Thought*, pp. 145–61.

28. Bachelard, *The Poetics of Space*, pp. 4, 5.

29. See the comments by architect Robert Mugerauer in his *Interpretations on Behalf of Place: Environmental Displacements and Alternative Responses*, New York: State University of New York Press 1994, especially Chapter 10.

30. A. J. Gurevich, *Medieval Popular Culture: Problems of Belief and Perception*, ET Cambridge: Cambridge University Press 1990, p. 79.

31. Mark R. Cohen, *Under Crescent and Cross: The Jews in the Middle Ages*, Princeton: Princeton University Press 1994, pp. 78–80.

32. David Canter, *The Psychology of Place*, London: The Architectural Press 1977, pp. 9–10, 158–9.

33. Penelope Lively, *Spiderweb*, Harmondsworth/New York: Penguin 1999, p. 13.

34. Cited in Julian Thomas, *Time, Culture and Identity*, London: Routledge 1999, p. 87.

35. See Bachelard, *The Poetics of Space*, p. 7.

36. Schama, *Landscape and Memory*, p. 61.

37. On the 'spirituality' implied by landscape art, not least the contrast between English landscape painting and the art of Australia and North America, see, for example, Peter Fuller, *Theoria: Art and the Absence of Grace*, London: Chatto and Windus 1988, especially chapters 14, 'An Earthly Paradise?', 19, 'The Art of England', and 21, 'The Glare of the Antipodes'.

38. W. G. Hoskins, *The Making of the English Landscape*, London: Hodder & Stoughton revised edn 1992, p. 18.

39. Kathleen Norris, *Dakota: A Spiritual Geography*, New York: Houghton Mifflin 1993, p. 2.

40. Belden Lane, Review of Schama, *Landscape and Memory*, *Christian Spirituality Bulletin* 4/1 (Summer 1996), p. 31.

41. Douglas Burton-Christie, 'Nature, Spirit and Imagination: The

Quest for Sacred Place', unpublished lecture delivered at The College of San Rafael, California, September 1995, p. 2.

42. Martyn Whittock, *Wiltshire Place-Names: Their Origins and Meanings*, Newbury: Countryside Books 1997, p. 122, and Eilert Ekwall, *The Concise Oxford Dictionary of English Place-Names*, Oxford: Clarendon Press 1985, p. 402.

43. Belden Lane, 'Galesville and Sinai: The Researcher as Participant in the Study of Spirituality and Sacred Space', *Christian Spirituality Bulletin* 2/1 (Spring 1994), p. 19.

44. André Brink, *Looking on Darkness*, London: Secker & Warburg 1982.

45. Paul Ricoeur, *Time and Narrative*, volume 1, ET Chicago: University of Chicago Press 1984, p. 3.

46. Paul Ricoeur, *Time and Narrative*, volume 3, ET Chicago: University of Chicago Press 1988, p. 103.

47. Ricoeur, *Time and Narrative*, 3, p. 202.

48. Ricoeur, *Time and Narrative*, 1, p. 75.

49. Mark Wallace, 'Introduction' to Paul Ricoeur, *Figuring the Sacred: Religions, Narrative and Imagination*, ET Minneapolis: Fortress Press 1995, p. 11.

50. See Ricoeur, *Time and Narrative*, especially volume 1, Part II 'History and Narrative'.

51. Schama, *Landscape and Memory*, pp. 67–71.

52. Henri Lefebvre, *The Production of Space*, ET Oxford: Blackwell 1991.

53. See, for example, the comments of Gustavo Gutiérrez in *We Drink from Our Own Wells*, ET London: SCM Press/New York: Orbis Books 1984, Part 1 and especially pp. 26–9.

54. For a summary of the impact of the 'new history' on studies of spirituality see Philip Sheldrake, *Spirituality and History: Questions of Interpretation and Method*, 2nd edn, London: SPCK 1995/New York: Orbis Books 1999, especially chapters 3 and 4.

55. See Elaine Graham, 'From Space to Woman-Space', *Feminist Theology* 9 (May 1995), pp. 11–34.

56. Cited by Allan Wolter in D. McElrath (ed.), *Franciscan Christology*, New York: Franciscan Institute 1980, pp. 141, 153.

57. In W. H. Gardner and N. H. MacKenzie (eds), *The Poems of Gerard Manley Hopkins*, Oxford: Oxford University Press 1989, p. 90.

58. 'Duns Scotus's Oxford', in Gardner and MacKenzie (eds), *The Poems of Gerard Manley Hopkins*, p. 79.

59. *Opus Oxoniense*, II, 3, 6, 2.

60. See Umberto Eco, *Art and Beauty in the Middle Ages*, New Haven: Yale University Press 1986, pp. 85–8.

61. For example, see Michael Blastic, 'Franciscan Spirituality', p. 416, in Michael Downey (ed.), *The New Dictionary of Catholic Spirituality*, Collegeville, MN: Liturgical Press 1993.

62. New translation from the medieval Italian by the contemporary Poor Clare scholar, Sister Frances Teresa. See her *Living the Incarnation: Praying with Francis and Clare of Assisi*, London: Darton, Longman & Todd 1993, p. 129. Used by permission of the publishers.

63. Translation in Regis Armstrong and Ignatius Brady (eds), *Francis and Clare: The Complete Works*, London: SPCK/New York: Paulist Press 1982, p. 154.

64. See, for example, Bronislaw Geremek, 'The Marginal Man', in Jacques Le Goff (ed.), *The Medieval World*, ET London: Collins and Brown 1990, especially pp. 367–9, and R. I. Moore, *The Formation of a Persecuting Society*, Oxford: Blackwell 1994, pp. 45–63.

65. In Michel de Certeau, 'How Is Christianity Thinkable Today?' ET in Graham Ward (ed.), *The Postmodern God*, Oxford: Blackwell 1997, p. 142.

66. De Certeau, 'How Is Christianity Thinkable Today?', p. 151.

2. Place in Christian Tradition

1. Augustine, *The City of God*, in *Nicene and Post-Nicene Fathers of the Christian Church*, volume II, Edinburgh: T&T Clark/Grand Rapids: Eerdmans 1993 edn, or the translation by Henry Bettenson, Harmondsworth: Penguin 1984.

2. See, for example, remarks in John McManners, 'Introduction', pp. 1–18, in John McManners (ed.), *The Oxford History of Christianity*, Oxford/New York: Oxford University Press 1993.

3. See John O'Meara, 'Introduction', p. vii, in Henry Bettenson, trans., *The City of God*, Harmondsworth: Penguin 1984.

4. See the classic work on Augustine's theory of history, R. A. Markus, *Saeculum: History and Society in the Theology of St Augustine*, Cambridge: Cambridge University Press 1970, especially Chapter 1 'History: Sacred and Secular'.

5. See Peter Brown, *Authority and the Sacred: Aspects of the Christianisation of the Roman World*, Cambridge: Cambridge University Press 1995, p. 60.

6. Peter Brown, *The Making of Late Antiquity*, Cambridge, MA: Harvard University Press, new edn 1993, Chapter 1 'A Debate on The Holy'.

7. See Brown, *Authority and the Sacred*, p. 74.

8. See John Wilkinson, trans., *Egeria's Travels*, London: SPCK 1971, 20,6.

9. See Life of Daniel, 7, in E. Dawes and N. Baynes, *Three Byzantine Saints*, Oxford: B. Blackwell 1948; and Brown, *The Making of Late Antiquity*, pp. 13–14.

10. To the modern observer, the old monastic sites on the Skelligs off the coast of Kerry or even on Iona off the West Coast of Scotland, appear dauntingly remote and inhospitable. In their own context, however, both places, while undoubtedly offering solitude, were also parts of networks of seaborne trade routes and of social and political connections.

11. On this distinction in relation to biblical texts, see Walter Brueggemann, *Interpretation and Obedience: From Faithful Reading to Faithful Living*, Minneapolis: Fortress Press 1991, especially p. 31.

12. A. J. Gurevich, *Medieval Popular Culture*, Cambridge: Cambridge University Press 1988, pp. 17–18.

13. See Paul Riceour, *Time and Narrative*, 3 vols, ET Chicago: University of Chicago Press 1984, 1985, 1988, especially volume 1, Part II 'History and Narrative'.

14. Alison Goddard Elliott, *Roads to Paradise: Reading the Lives of the Early Saints*, Hanover: University Press of New England 1987, Chapter 1 'Introduction'.

15. See, for example, Philip Sheldrake, *Spirituality and History: Questions of Interpretation and Method*, 2nd edn, London: SPCK 1995/ New York: Orbis 1998, pp. 105–7.

16. Elliott, *Roads to Paradise*, p. 6. The whole book is a fascinating study of the narratives of early Christian saints and their utopian vision.

17. Elliott, *Roads to Paradise*, pp. 90–1, and the article by Peter Brown, 'The Rise and Function of the Holy Man in Late Antiquity', in his *Society and the Holy in Late Antiquity*, Berkeley: University of California Press 1989, pp. 103–52.

18. See the Coptic *The Life of Onnophrius*, 2, ET in Tim Vivian (ed.), *Journeying into God: Seven Early Monastic Lives*, Minneapolis: Fortress Press 1996. The ms dates from the tenth century but Onnophrius has been dated possibly to the mid-fourth century. The name Paphnutius, 'servant of God' in Coptic, may or may not refer to a single historical figure. For comments on the text see Elliott, *Roads to Paradise*, pp. 51–8.

19. For a scholarly English translation from the Spanish Autograph, see George Ganss (ed.), *The Spiritual Exercises of Saint Ignatius*, Chicago: Loyola University Press 1992. The sections on the criteria for choosing appear in 'The Principle and Foundation' at section 23 (Exx 23 in standard contemporary numeration) and in the process of 'Making an Election' (or choice) at Exx 169–89.

20. Brown, *Authority and the Sacred*, p. 62.

21. Gurevich, *Medieval Popular Culture*, p. 205.

22. Donald Weinstein and Rudolph M. Bell, *Saints and Society: The Two Worlds of Western Christendom 1000–1700*, Chicago: University of Chicago Press 1982, pp. 176–7.

23. André Vauchez, *Sainthood in the Later Middle Ages*, ET Cambridge: Cambridge University Press 1997, p. 133.

24. Weinstein and Bell, *Saints and Society*, pp. 186–7.

25. See Weinstein and Bell, *Saints and Society*, Part 2, 'Perceptions of Sanctity', Chapter 6 'Place'.

26. For example, Weinstein and Bell, *Saints and Society*; John Stratton Hawley (ed.), *Saints and Virtues*, Berkeley: University of California Press 1987; Renate Blumenfeld-Kosinski and Timea Szell (eds), *Images of Sainthood in Medieval Europe*, Ithaca: Cornell University Press 1991.

27. Sheldrake, *Spirituality and History*, Chapter 3, *passim*.

28. St Bonaventure suggested that St Francis was attending a Mass of the Apostles. See his Life of St Francis (*Legenda Maior*), ET in Ewert Cousins (ed.), *Bonaventure: The Soul's Journey into God; The Tree of Life; The Life of St Francis*, Classics of Western Spirituality, New York: Paulist Press 1978, Chapter 3 'On the Foundation of the Order and the Approval of the Rule', pp. 199–200.

29. The text of the *Later Rule* is from Chapter 6, 2, in Regis Armstrong and Ignatius Brady (eds), *Francis and Clare: The Complete Works*, London: SPCK/New York: Paulist Press 1982. See also Jacques le Goff, 'Francis of Assisi between the Renewals and Restraints of Feudal Society', *Concilium* 149 (1981), pp. 3ff., and Hester Goodenough Gelber, 'A Theater of Virtue: The Exemplary World of St Francis of Assisi', in John Stratton Hawley, *Saints and Virtues*, Berkeley: University of California Press 1987, pp. 15ff.

30. See Weinstein and Bell, *Saints and Society*, Chapter 7 'Class', and Michael Goodich, '*Ancilla Dei*: The Servant as Saint in the Late Middle Ages', in J. Kirshner and S. Wemple (eds), *Women of the Medieval World*, Oxford: Oxford University Press 1987, pp. 119ff.

31. See Peter Brown, *The Cult of the Saints: Its Rise and Function in*

Latin Christianity, London: SCM Press/Chicago: University of Chicago Press 1981, Chapter 1 'The Holy and The Grave'.

32. Cited in Brown, *The Cult of the Saints*, p. 8 n. 33.

33. Second Epistle, especially 7–15, PG 46 1013B, cited in Simon Coleman and John Elsner, *Pilgrimage*, London: British Museum Press/Cambridge, MA: Harvard University Press 1995, pp. 80–1. For Gregory's apophatic perspective on place, see Jaroslav Pelikan, *Christianity and Classical Culture*, New Haven: Yale University Press 1993, p. 113.

34. Cited in Gurevich, *Medieval Popular Culture*, p. 40.

35. Cited in Gurevich, *Medieval Popular Culture*, pp. 40–2.

36. Gurevich, *Medieval Popular Culture*, p. 42.

37. Albert Rouet, *Liturgy and the Arts*, ET Collegeville, MN: Liturgical Press 1997, p. 95.

38. Rouet, *Liturgy and the Arts*, p. 105.

39. Umberto Eco, *Art and Beauty in the Middle Ages*, ET New Haven: Yale University Press 1986.

40. For some reflections on what might be called the theology of Gothic, see Christopher Wilson, *The Gothic Cathedral*, London: Thames & Hudson 1990, especially the Introduction, pp. 64–6, 219–20, 262–3.

41. On this mixture of theological aesthetics see the essay by Bernard McGinn, 'From Admirable Tabernacle to the House of God: Some Theological Reflections on Medieval Architectural Integration', in Virginia Chieffo Raguin, Kathryn Brush and Peter Draper (eds), *Artistic Integration in Gothic Buildings*, Toronto: University of Toronto Press 1995.

42. A significant theological essay on Gothic cathedrals is McGinn, 'From Admirable Tabernacle to the House of God', pp. 41–56.

43. *Libellus Alter De Consecratione Ecclesiae Sancti Dionysii*, IV, trans. in Erwin Panofsky, *Abbot Suger on the Abbey Church of St Denis and Its Art Treasures*, Princeton: Princeton University Press 1979, pp. 100–1.

44. ET in Philip Schaff (ed.), *A Select Library of the Nicene and Post-Nicene Fathers of the Christian Church*, Michigan: Eerdmans/Edinburgh: T&T Clark 1996 edn, volume VIII St Augustine 'Expositions on the Book of Psalms', Psalm XLII, section 8, p. 134.

45. Translation by Bernard McGinn cited in his 'From Admirable Tabernacle to the House of God', p. 49, from the Latin text in Panofsky, *Abbot Suger*, p. 104.

46. On such practicalities and their impact on architectural developments, see Richard Morris, *Churches in the Landscape*, London: Dent 1989, pp. 96–8 and 289–95.

47. *The Divine Names*, Chapter 4.4, ET in Colm Luibheid (ed.), *Pseudo-Dionysius: The Complete Works*, London: SPCK 1987, p. 74.

48. For a brief summary of Dionysian theory, see Sheldrake, *Spirituality and History*, pp. 200–1. Also Georges Duby, *The Age of the Cathedral: Art and Society 980–1420*, ET Chicago: University of Chicago Press 1981, pp. 99–100.

49. *The Mystical Theology*, Chapter 1.1, ET in Luibheid, *Pseudo-Dionysius*, p. 135.

50. Michael Camille, *Gothic Art: Visions and Revelations of the Medieval World*, London: Weidenfeld & Nicolson 1996, p. 12. Camille is one of the most commended representatives of the new generation of scholars of medieval art.

51. In *De Consecratione*, translated in Panofsky, *Abbot Suger*, p. 82.

52. See Colleen McDannell and Bernhard Lang, *Heaven: A History*, New Haven: Yale University Press 1988, pp. 70–80. Also Duby, *The Age of the Cathedral*, Part 2.

53. Quoted in translation in McDannell and Lang, *Heaven: A History*, p. 79.

54. Duby, *The Age of the Cathedral*, p. 95.

55. Aloysius Pieris, 'Spirituality and Liberation', *The Month* (April 1983), p. 120.

56. Otto von Simson, *The Gothic Cathedral: Origins of Gothic Architecture and the Medieval Concept of Order*, Princeton: Princeton University Press, expanded edn 1989.

57. See Franz Leenhardt, *Two Biblical Faiths: Protestant and Catholic*, ET London: SCM Press 1964, and Hieje Faber, *Above the Treeline: Towards a Contemporary Spirituality*, London: SCM Press 1988.

58. Wolfhart Pannenberg, *Christian Spirituality and Sacramental Community*, ET London: SCM Press 1984, pp. 15–22.

59. Rudolph Bultmann, *Jesus Christ and Mythology*, New York: Scribners 1958, pp. 84–5.

60. John Calvin, *Institutes of The Christian Religion*, trans. Henry Beveridge, repr. Grand Rapids: Eerdmans 1995, I, XIV, 20.

61. Rowan Williams, 'Sacraments of the New Society', in David Brown and Ann Loades (eds), *Christ: The Sacramental Word*, London: SPCK 1996, pp. 89–90.

62. See contrasting essays by Professor Susan White and former Archbishop John Habgood in David Brown and Ann Loades (eds), *The Sense of the Sacramental: Movement and Measure, Art and Music, Place and Time*, London: SPCK 1995.

3. The Eucharist and Practising Catholic Place

1. Richard McBrien, *Catholicism*, New York: HarperCollins 1994, p. 7.

2. For a detailed analysis of 'thisness', see Allan B. Wolter, *The Philosophical Theology of John Duns Scotus*, Ithaca, New York: Cornell University Press 1990, pp. 89–97.

3. See the contrasting essays by Susan White and John Habgood in David Brown and Ann Loades (eds), *The Sense of the Sacramental: Movement and Measure in Art and Music, Place and Time*, London: SPCK 1995, pp. 31–43 and 19–30 respectively. White's ethical approach needs to be balanced by Habgood's more markedly sacramental understanding of the whole of the natural world.

4. See *Summa Theologiae* 1.8.1.

5. See David Tracy, 'The Return of God in Contemporary Theology', *Concilium* 94 no. 6, *Why Theology?*, pp. 37–46.

6. See David Tracy, *On Naming the Present: God, Hermeneutics, and Church*, New York: Orbis Books 1994, pp. 42–5.

7. Karl Rahner, *The Trinity*, ET London: Burns & Oates 1970, p. 22.

8. For the notion of space in God, see Colin Gunton, *The Promise of Trinitarian Theology*, Edinburgh: T&T Clark 1997, pp. 112ff. See also his *The One, The Three and The Many*, Cambridge: Cambridge University Press 1995, p. 164 where he writes, 'There is thus a richness and space in the divine life, in itself and as turning outwards in the creation of the dynamic universe that is relational order in space and time.'

9. On reconceiving particularity in Trinitarian terms, see David S. Cunningham, *These Three Are One: The Practice of Trinitarian Theology*, Oxford: Blackwell 1998, Chapter 6.

10. See Gunton, *The One, The Three and The Many*, p. 113.

11. Belden C. Lane, *The Solace of Fierce Landscapes: Exploring Desert and Mountain Spirituality*, New York: Oxford University Press 1998, p. 46.

12. Rowan Williams and Philip Sheldrake, 'Catholic Persons: Images of Holiness. A Dialogue', in Jeffrey John (ed.), *Living the Mystery*, London: Darton, Longman & Todd 1994, pp. 76–8.

13. Avery Dulles, *The Catholicity of the Church*, Oxford: Clarendon Press 1985.

14. McBrien, *Catholicism*, pp. 8–16.

15. *Exercises*, para. 236 ET in George Ganss (ed.), *The Spiritual Exercises of Saint Ignatius*, Chicago: Loyola University Press 1992.

16. Thomas Traherne, *Centuries*, London: Mowbray 1975, 1, 31.

17. Charles Curran, *The Church and Morality: An Ecumenical and Catholic Approach*, Minneapolis: Fortress Press 1993, pp. 18–20.

18. For an interesting summary of these contrasting visions of Catholic identity, see Norbert Greinacher, 'Catholic Identity in the Third Epoch of Church History', in James Prevost and Knut Walf (eds), *Concilium* 1994 no. 5, *Catholic Identity*, pp. 3–14.

19. Curran, *The Church and Morality*, p. 9.

20. The link between the enactment of identity and the ethical nature of the Eucharist is discussed by the contemporary moral theologian, William Spohn, *Go and Do Likewise: Jesus and Ethics*, New York: Continuum 1999, pp. 175–84.

21. On this point, see Donald E. Saliers, 'Liturgy and Ethics: Some New Beginnings', in Ronald Hamel and Kenneth Himes (eds), *Introduction to Christian Ethics: A Reader*, New York: Paulist Press 1989, pp. 175–86.

22. *Baptism, Eucharist and Ministry*, Faith and Order Paper 111, Geneva: World Council of Churches 1982, paras 19–20 and 22.

23. Rowan Williams, *On Christian Theology*, Oxford: Blackwell 2000, pp. 209–10.

24. See William Cavanaugh, 'The Eucharist as Resistance to Globalisation', in Sarah Beckwith (ed.), *Catholicism and Catholicity: Eucharistic Communities in Historical and Contemporary Perspectives*, Oxford: Blackwell 1999, pp. 69–84.

25. Michel de Certeau, *The Practice of Everyday Life*, ET Berkeley: University of California Press 1988, pp. 115–30.

26. See Tracy, *On Naming the Present*, pp. 31–4.

27. *Baptism, Eucharist and Ministry*, para. 24.

28. David Ford, *Self and Salvation*, Cambridge: Cambridge University Press 1999, touches on what he calls 'the ethics of feasting', pp. 268–70.

29. See Christopher Rowland, 'Eucharist as Liberation from The Present', in Brown and Loades, *The Sense of the Sacramental*, pp. 200–15.

30. For a radical social reading of the sacraments, especially the Eucharist, see Rowan Williams, *On Christian Theology*, Oxford: Blackwell 2000, Chapter 14 'Sacraments of the New Society'.

31. The classic work remains Avery Dulles, *Models of the Church*, New York: Image Books 1987.

32. William Cavanaugh, *Torture and the Eucharist: Theology, Politics and the Body of Christ*, Oxford: Blackwell 1998, pp. 207–21.

33. Williams, *On Christian Theology*, pp. 212–14.

34. *Baptism, Eucharist and Ministry*, para. 14.

35. Victor Codina, 'Sacraments', pp. 218–19, in Jon Sobrino and Ignacio Ellacuria (eds), *Systematic Theology: Perspectives from Liberation Theology*, ET London: SCM Press 1996.

36. See Graham Ward, 'The Displaced Body of Jesus Christ', in John Milbank, Catherine Pickstock and Graham Ward (eds), *Radical Orthodoxy*, London: Routledge 1999, pp. 163–81.

37. See Codina, 'Sacraments', p. 228.

38. Ignacio Ellacuria, 'The Church of the Poor, Historical Sacrament of Liberation', in Ignacio Ellacuria and Jon Sobrino (eds), *Mysterium Liberationis: Foundational Concepts of Liberation Theology*, ET New York: Orbis Books 1993, p. 543.

39. On corporeality and the incorporation of the Church in history, see Ellacuria, 'The Church of the Poor', p. 545.

40. On the prophetic quality of Jesus' incorporation, see Ford, *Self and Salvation*, pp. 151–2.

41. Catherine Keller, *Apocalypse Now and Then*, Boston: Beacon Press 1996, pp. 142–3.

42. See Cavanaugh, *Torture and the Eucharist*, pp. 11–18.

43. Ford, *Self and Salvation*, p. 163. Although Ford's precise phrase concerns baptism, his wider context is the theology of the Eucharist.

44. *Baptism, Eucharist and Ministry*, para. 18.

4. The Practice of Place: Monasteries and Utopias

1. Robin Gill, 'Churches as Moral Communities', in *Moral Communities: The Prideaux Lectures 1992*, Exeter: Exeter University Press 1992, pp. 63–80.

2. Richard Morris, *Churches in the Landscape*, London: Dent 1989, p. 104.

3. Andrew Louth in *The Wilderness of God*, London: Darton, Longman & Todd 1991, represents a recent attempt to describe the special qualities of the religion of the desert. The essay 'The Wilderness in the Medieval West' by Jacques Le Goff in his *The Medieval Imagination*, ET London/Chicago: University of Chicago Press 1988, has some illuminating remarks on the understanding of 'desert' in Western monasticism, including the Celtic tradition.

4. Robert C. Gregg (ed.), *Athanasius: The Life of Anthony*, New York: Paulist Press 1980, para. 8, pp. 37–9.

5. See Peter Brown, *The Making of Late Antiquity*, Cambridge, MA: Harvard University Press, new edn 1993, Chapter 4 'From the Heavens to the Desert: Anthony and Pachomius'.

6. On the geographics of monastic 'style', see Peter Brown, *Society and the Holy in Late Antiquity*, Berkeley: University of California Press 1989, pp. 110–14.

7. George Lawless, *Augustine of Hippo and His Monastic Rule*, Oxford: Clarendon Press 1987, 'Regulations for a Monastery', Chapter 4, 4, p. 89. Cited as 'Rule'.

8. See, for example, the experience of monastic space as paradise restored in 'The Life of St Onophrius', ET in Tim Vivian (ed.), *Journeying into God: Seven Early Monastic Lives*, Minneapolis: Fortress Press 1996.

9. See, for example, Jacques Ellul, *The Meaning of the City*, ET Carlisle: Paternoster Press 1997.

10. On this point, see Peter Brown, 'The Rise and Function of the Holy Man in Late Antiquity', in his *Society and the Holy in Late Antiquity*, Berkeley: University of California Press 1989.

11. On More's Utopia, see Edward Surtz and J. H. Hexter (eds), *Utopia*, The Complete Works of St Thomas More, volume 4, New Haven: Yale University Press 1965.

12. John Carey, *The Faber Book of Utopias*, London: Faber and Faber 1999, p. xxvi.

13. See Henri Lefebvre's comments on utopias in his philosophical study of 'daily life', urbanism and architecture, *The Production of Space*, ET Oxford: Blackwell 1991, p. 60.

14. Lefebvre, *The Production of Space*, pp. 163–4.

15. For an illuminating essay on the nature of utopias, see Professor Carey's Introduction in *The Faber Book of Utopias*, pp. xi–xxvi.

16. Carol Lake, *Rosehill: Portraits from a Midlands City*, London: Bloomsbury 1989.

17. See Surtz and Hexter, *Utopia*, Introduction, pp. xlviii, lxxv–lxxvii.

18. Surtz and Hexter, *Utopia*, Introduction, p. lxxvii.

19. For some illuminating remarks on Renaissance rhetoric, particularly in reference to the connections between rhetoric and place, see Marjorie O'Rourke Boyle, *Loyola's Acts: The Rhetoric of the Self*, Berkeley: University of California Press 1997, especially pp. 7–10.

20. Benedicta Ward (ed.), *The Wisdom of the Desert Fathers*, Oxford: Fairacres Publications 1986, no. 70.

21. For an illuminating study of mountains as symbols of the spiritual journey in the Christian tradition, see Chapter 5 'Sinai and Tabor:

Mountain Symbolism in the Christian Tradition', in Belden Lane, *The Solace of Fierce Landscapes: Exploring Desert and Mountain Spirituality*, New York: Oxford University Press 1998.

22. See Lane, *The Solace of Fierce Landscapes*, pp. 141–7.

23. See Marjorie Reeves, *Joachim of Fiore and the Prophetic Future*, London: SPCK 1976.

24. ET in Bernard McGinn (ed.), *Apocalyptic Spirituality – Treatises and Letters of Lactantius, Adso of Montier-en-Der, Joachim of Fiore, The Spiritual Franciscans, Savanarola*, New York: Paulist Press 1979, pp. 142–8.

25. See, for example, Michel Foucault, *Discipline and Punish: The Birth of the Prison*, ET Harmondsworth: Penguin 1991 (originally published in 1975).

26. The validity of theological interpretations of a post-Christian (and probably atheist) thinker, albeit one fascinated by the history of Christianity and, in later years, by theology, is discussed intelligently by his main English-speaking interpreter, Jeremy R. Carrette in his *Foucault and Religion: Spiritual Corporality and Political Spirituality*, London/New York: Routledge 2000, pp. ix–xii, 1–6. There is a critical appraisal of Foucault's writing on monasticism on pp. 112–13, 118–22.

27. Quoted in Katherine Gibson and Sophie Watson, 'An Introduction', in Katherine Gibson and Sophie Watson (eds), *Postmodern Cities and Spaces*, Oxford: Blackwell 1995, p. 2.

28. Michel Foucault, 'Of Other Spaces', 1967, quoted in Edward W. Soja, 'Heterotopologies: A Remembrance of Other Spaces in the Citadel-LA', in Gibson and Watson, *Postmodern Cities and Spaces*, p. 14.

29. For example, Philip Sheldrake, *Living between Worlds: Place and Journey in Celtic Spirituality*, London: Darton, Longman & Todd/Boston: Cowley 1995, pp. 22–5.

30. See Graham Ward (ed.), *The Certeau Reader*, Oxford: Blackwell 2000, 'Introduction', pp. 1–14.

31. 'Walking in the City', pp. 103–4, in Ward, *Certeau Reader*.

32. See de Certeau's essay, 'The Weakness of Believing. From the Body to Writing, a Christian Transit', ET in Ward, *Certeau Reader*, p. 221.

33. Michel de Certeau, 'Culture and Spiritual Experience', in *Concilium* 19 (1966), p. 22.

34. de Certeau, 'The Weakness of Believing', p. 226.

35. de Certeau, 'The Weakness of Believing', p. 236.

36. de Certeau, 'The Weakness of Believing', p. 215.

37. See Umberto Eco, *Serendipities: Language and Lunacy*, London: Weidenfeld & Nicolson/New York: Columbia University Press 1999, especially chapters 2 and 4.

38. On the different interpretations of silence in monastic writers and among those leading the canonical life, see Caroline Walker Bynum, *Jesus as Mother: Studies in the Spirituality of the High Middle Ages*, Berkeley: University of California Press 1984, Chapter 1.

39. Peter of Porto, *Regula clericorum*, Book 1, chapters 32–6, in Migne, *Patrologia Latina* 163, columns 720–2, cited in Bynum, *Jesus as Mother*, p. 45.

40. Benedicta Ward, *Wisdom of the Desert Fathers*, no. 106.

41. Benedicta Ward, *Wisdom of the Desert Fathers*, no. 108.

42. G. S. M. Walker (ed.), *Sancti Columbani Opera*, Scriptores Latini Hiberniae, volume II, Dublin: The Dublin Institute for Advanced Studies 1970, pp. 124–5.

43. For an excellent study of the theory and practice of speech in early monasticism, see Douglas Burton-Christie, *The Word in The Desert: Scripture and the Quest for Holiness in Early Christian Monasticism*, New York: Oxford University Press 1993, Chapter 5 'Words and Praxis'.

44. *The Rule of The Master*, Chapter L, verse 13. ET Kalamazoo: Cistercian Publications 1977.

45. Giovanni Miccoli, 'Monks', in Jacques Le Goff (ed.), *The Medieval World*, London: Collins & Brown 1990, p. 43.

46. *Summa Theologiae* I. II. 4:8.

47. Giles of Rome, *Quodlibeta*, 6:25, quoted in Colleen McDannell and Bernhard Lang, *Heaven: A History*, New Haven: Yale University Press 1988, p. 93.

48. This is the emphasis of Augustine's commentary on Genesis and is cited in Robert Markus, *The End of Ancient Christianity*, Cambridge: Cambridge University Press 1990, p. 78.

49. Reference to Acts 4.32. See 'Rule', 4, p. 75, in Lawless, *Augustine of Hippo*.

50. Lawless, *Augustine of Hippo*, 'Rule', Chapter 1, 3, p. 81.

51. Lawless, *Augustine of Hippo*, 'Rule', Chapter 5, 2, p. 95.

52. See Timothy Fry OSB (ed.), *The Rule of St. Benedict in Latin and English with Notes*, Collegeville, MN: Liturgical Press 1981.

53. See, for example, Peter Brown, 'The Notion of Virginity in the Early Church', in Bernard McGinn and John Meyendorff (eds), *Christian Spirituality: Origins to the Twelfth Century*, New York: Crossroad 1986, and Brown's more developed ideas in *The Body and Society: Men, Women*

and Sexual Renunciation in Early Christianity, London: Faber and Faber 1991.

54. See, for example, Jo Ann Kay McNamara, 'Introduction: Chastity and Female Identity' *Sisters in Arms: Catholic Nuns through Two Millennia*, Cambridge, MA: Harvard University Press 1996.

55. Benedicta Ward, *Wisdom of the Desert Fathers*, nos 72, 74, 68, 63.

56. Lawless, *Augustine of Hippo*, 'Rule', Chapter 1, 6 and 7, p. 83.

57. See Lisa Bitel, *Isle of the Saints: Monastic Settlement and Christian Community in Early Ireland*, Ithaca, New York: Cornell University Press 1990, p. 82.

58. See Kathleen Hughes and Ann Hamlin, *Celtic Monasticism: The Modern Traveller to the Irish Church*, New York: Seabury Press 1981, pp. 13–15, 54; also Kathleen Hughes, *Church and Society in Ireland AD 400–1200*, London: Variorum 1987, section 8 'The Church and the World in Early Christian Ireland', p. 111.

59. See, for example, Benedicta Ward, *Wisdom of the Desert Fathers*, no. 202. Similar stories are attributed to contemporary monastic life in Ethiopia in 'Ethiopian Orthodox Models of Religious Life', a privately circulated document made available by Sister Elizabeth Rees at the Ammerdown Centre near Bath.

60. For a historical and theological reflection on the Christian tradition of hospitality, see Christine D. Pohl, *Making Room: Recovering Hospitality as a Christian Tradition*, Grand Rapids: Eerdmans 1999.

61. Benedicta Ward, *Wisdom of the Desert Fathers*, nos 151, 224.

62. Bitel, *Isle of the Saints*, pp. 201–2; Hughes and Hamlin, *Celtic Monasticism*, pp. 14, 75.

63. See Marjorie O'Rourke Boyle, *Divine Domesticity: Augustine of Thagaste to Teresa of Avila*, Leiden: Brill 1997.

64. For example, commentary on Psalm 122 in Augustine's *Homilies on the Psalms*. See Mary T. Clark (ed.), *Augustine of Hippo: Selected Writings*, London: SPCK/New York: Paulist Press 1984, pp. 249–50.

65. See Philip Sheldrake, *Spirituality and History: Questions of Interpretation and Method*, 2nd edn, London: SPCK 1995/New York: Orbis 1999, pp. 118–19.

66. Peter Brown, *The Body and Society: Men, Women and Sexual Renunciation in Early Christianity*, London: Faber and Faber 1991, p. 227.

67. *Sancti Columbani Opera*, Sermon VIII, 2, p. 97.

68. Quoted in Nora Chadwick, *The Age of the Saints in the Early Celtic Church*, Oxford: Oxford University Press 1961, p. 83.

69. *Sancti Columbani Opera*, Sermon VIII, 2; p. 97 ll 11–13.

70. See Chadwick, *The Age of the Saints*, pp. 30–2. Also Peter Harbison, *Pilgrimage in Ireland*, London: Barrie & Jenkins 1991, pp. 33–4.

71. *Later Rule*, chapter 6.2. ET in Regis Armstrong and Ignatius Brady (eds), *Francis and Clare: The Complete Works*, London: SPCK/New York: Paulist Press 1982.

72. For a contemporary scholarly edition in English, see George E. Ganss (ed.), *The Constitutions of the Society of Jesus*, St Louis: The Institute of Jesuit Sources 1970. Subsequent references to both the *Formula* and the *Constitutions* follow the standard modern paragraph numbers.

73. See, for example, Chapter II 'Ignatius and the Ascetic Tradition of the Fathers', in Hugo Rahner, *Ignatius the Theologian*, ET London: Geoffrey Chapman 1990.

74. For the most up-to-date scholarship on the foundation of the Society of Jesus, and interpretation of the contrasts with traditional monasticism, see John O'Malley, *The First Jesuits*, Boston, MA: Harvard University Press 1993.

75. *The First and General Examen which should be proposed to all who request admission into the Society of Jesus*, 7 in Ganss, *The Constitutions of the Society of Jesus*, pp. 79–80. See also *Formula of the Institute*, 4 in Ganss, p. 68, and *Constitutions*, 529 and 603, in Ganss, pp. 239 and 268. For commentaries on the role of mobility in the *Constitutions*, see Joseph Veale, 'How the Constitutions Work', pp. 6, 9, 10, and Howard Gray, 'What Kind of Document?', pp. 24–5, and Brian O'Leary, 'Living with Tension', pp. 39–40, all in *The Way Supplement* 61 (Spring 1988).

76. Jerome Nadal, *Monumenta Historica Societatis Jesu, Monumenta Nadal*, V nos 195, 773.

77. On Nadal's theology of nature and grace, see O'Malley, *The First Jesuits*, pp. 68, 214, 242, 250, 281–2.

78. *The Spiritual Exercises*, nos 101–9. There are several contemporary English translations, for example the one in the Classics of Western Spirituality series: George Ganss (ed.), *Ignatius of Loyola: Spiritual Exercises and Selected Works*, New York: Paulist Press 1991.

5. The Mystical Way: Transcending Places of Limit

1. Bernard McGinn, *The Foundations of Mysticism: Origins to the Fifth Century*, 'The Presence of God. A History of Western Christian Mysticism', volume 1, New York: Crossroad 1992, p. xvi.

2. See, for example, Karl Rahner, 'The Theology of Mysticism', in

K. Lehmann and I.. Raffelt (eds), *The Practice of Faith: A Handbook of Contemporary Spirituality*, ET New York: Crossroad 1986, pp. 70–7; Rowan Williams, *Teresa of Avila*, London: Geoffrey Chapman 1991, Chapter 5.

3. Rowan Williams, 'The Via Negativa and the Foundations of Theology: An Introduction to the Thought of V. N. Lossky', in Stephen Sykes and Derek Holmes (eds), *New Studies in Theology* 1, London: Duckworth 1980, p. 96.

4. See David Tracy, *The Analogical Imagination: Christian Theology and the Culture of Pluralism*, London: SCM Press 1981, repr. New York: Crossroad 1991, pp. 360, 385.

5. Denys Turner, *The Darkness of God: Negativity in Christian Mysticism*, Cambridge: Cambridge University Press 1995, p. 7.

6. On the patristic meaning of mysticism, see, for example, Louis Bouyer, 'Mysticism – An Essay on the History of the Word', in Richard Woods (ed.), *Understanding Mysticism*, London: Athlone Press 1981, pp. 42–55.

7. Turner, *The Darkness of God*, p. 262.

8. Turner, *The Darkness of God*, pp. 1–8.

9. Turner, *The Darkness of God*, p. 8.

10. Turner, *The Darkness of God*, p. 264.

11. Turner, *The Darkness of God*, p. 252.

12. Turner, *The Darkness of God*, p. 253.

13. See, for example, Thomas Merton, *Faith and Violence*, Notre Dame: University of Notre Dame Press 1968, *Conjectures of a Guilty Bystander*, London: Sheldon Press 1977, *Raids on the Unspeakable*, New York: New Directions 1966.

14. See, for example, the criticisms of definitions of the mystical as subjective experiences in Bernard McGinn, *The Foundations of Mysticism*, volume 1, General Introduction, pp. xi–xx, and Turner, *The Darkness of God*, Introduction, pp. 1–8.

15. Caroline Walker Bynum, *Jesus as Mother: Studies in the Spirituality of the High Middle Ages*, Berkeley: University of California Press 1982, Chapter V 'Women Mystics in the Thirteenth Century: The Case of the Nuns of Helfta'.

16. See, for example, Philip Sheldrake, *Spirituality and History: Questions of Interpretation and Method*, 2nd edn, London: SPCK 1995/ New York: Orbis Books 1998, pp. 79–80, 82, 127–8, 151, 155, 157 and the associated bibliographical references.

17. Grace Jantzen, *Power, Gender and Christian Mysticism*, Cambridge:

Cambridge University Press 1995, Chapter 5 ' "Cry Out and Write": Mysticism and the Struggle for Authority'.

18. For example, Nicholas Lash, 'The Church in the State We're in', in L. Gregory Jones and James J. Buckley (eds), *Spirituality and Social Embodiment*, Oxford: Blackwell 1997, p. 126.

19. John Ruusbroec, 'The Sparkling Stone', Conclusion, ET in James Wiseman (ed.), *John Ruusbroec: The Spiritual Espousals and Other Works*, New York: Paulist Press 1985, p. 184.

20. Chapter XVII. For an ET see James Walsh (ed.), *The Cloud of Unknowing*, New York: Paulist Press 1981.

21. John Ruusbroec, *Werken*, volume III, ET quoted in Paul Verdeyen, *Ruusbroec and His Mysticism*, Collegeville, MN: Liturgical Press 1994, p. 116.

22. Ruusbroec, *The Spiritual Espousals*, ET quoted in Verdeyen, *Ruusbroec and His Mysticism*, pp. 116–17.

23. *The Spiritual Espousals*, Book II, 'The Interior Life', in Wiseman (ed.), *John Ruusbroec*, pp. 136–43.

24. Evelyn Underhill, *Mysticism: The Nature and Development of Spiritual Consciousness*, 1911, repr. Oxford: One World Publications 1993.

25. Friedrich von Hügel, *The Mystical Element of Religion*, 1923, repr. London: James Clark 1927, vol. 2, pp. 264–5, 290–3 and 305–9.

26. Underhill, *Mysticism*, p. 172. On this point see pp. 172–4.

27. Underhill, *Mysticism*, p. 304.

28. Ruusbroec, *The Spiritual Espousals*, Book III, part 4.

29. Ruusbroec, *The Spiritual Espousals*, Book I, part 2, chap. lxv (Underhill, *Mysticism*, p. 436).

30. *The Spiritual Exercises*, para. 236, ET in George Ganss (ed.), *The Spiritual Exercises of Saint Ignatius*, Chicago: Loyola University Press 1992.

31. De Certeau, 'Mystic Speech', ET in Graham Ward (ed.), *The Certeau Reader*, Oxford: Blackwell 2000, p. 191.

32. De Certeau, 'Mystic Speeech', pp. 191–2.

33. See especially Jantzen, *Power, Gender and Christian Mysticism*, Chapter 1 'Feminists, Philosophers and Mystics'.

34. For example, Grace Jantzen, *Julian of Norwich*, London: SPCK 1987, p. 11, and her *Power, Gender and Christian Mysticism*, p. 178. Also Frederick Christian Bauerschmidt, *Julian of Norwich and the Mystical Body Politic of Christ*, Notre Dame: University of Notre Dame Press 1999, pp. 38, 75 (and n. 33); Joan Nuth, *Wisdom's Daughter: The Theology of Julian of Norwich*, New York: Crossroad 1991, pp. 16–22.

35. See, for example, the remarks of Ann Candlin in 'Reflections on Elaine Graham's Model of Space', *Feminist Theology* 9 (May 1995), pp. 82–95. This critical point seems to me to be valid even though Candlin relies on highly unreliable and selective translations of the mystics and takes considerable historical liberties in her brief accounts of the mystics!

36. Edmund Colledge and James Walsh (eds), *Julian of Norwich: Showings*, New York: Paulist Press 1978, Short Text, Chapter vi, p. 135.

37. Colledge and Walsh, *Julian of Norwich*, p. 177.

38. Bauerschmidt, *Julian of Norwich*.

39. See, for example, Paul Molinari, *Julian of Norwich: The Teaching of a Fourteenth Century English Mystic*, London: Burns & Oates 1958.

40. For example, Edmund Colledge and James Walsh (eds), *A Book of Showings to the Anchoress Julian of Norwich*, Montreal: Pontifical Institute of Mediaeval Studies 1978, Part 1, pp. 52, 72; Bauerschmidt, *Julian of Norwich*, p. 34; Denise Nowakowski Baker, *Julian of Norwich's Showings*, Princeton: Princeton University Press 1994, Chapter 1.

41. For example, Ewert Cousins, 'The Humanity and Passion of Christ', in Jill Raitt (ed.), *Christian Spirituality II: High Middle Ages and Reformation*, London: Routledge & Kegan Paul 1987; SCM Press 1989, p. 387, and Baker, *Julian of Norwich's Showings*, p. 47.

42. Cousins, 'The Humanity and Passion of Christ', p. 383.

43. See Ewert Cousins, 'Franciscan Roots of Ignatian Mysticism', in George Schner (ed.), *Ignatian Spirituality in a Secular Age*, Waterloo, Ontario: Wilfred Laurier University Press 1984, p. 60.

44. Jantzen, *Power, Gender and Christian Mysticism*, p. 346.

45. David Tracy, *On Naming the Present: God, Hermeneutics and Church*, New York: Orbis Books 1994, pp. 3–6.

46. Michel Foucault, *Power/Knowledge: Selected Interviews and Other Writings 1972–77*, ET London: Pantheon Books 1980, p. 81.

47. Segundo Galilea, 'The Spirituality of Liberation', *The Way* (July 1985), pp. 186–94.

48. Segundo Galilea, 'Liberation as an Encounter with Politics and Contemplation', ET in Richard Woods (ed.), *Understanding Mysticism*, London: Athlone Press 1981, p. 531.

49. Galilea, 'Liberation as an Encounter', pp. 535, 536.

50. Leonardo Boff, 'The Need for Political Saints: From a Spirituality of Liberation to the Practice of Liberation', ET in *Cross Currents* XXX/4 (Winter 1980/81), p. 371.

51. Boff, 'The Need for Political Saints', p. 373.

52. Boff, 'The Need for Political Saints', p. 372.

53. Boff, 'The Need for Political Saints', p. 374.

54. Jürgen Moltmann, 'The Theology of Mystical Experience', *Experiences of God*, Philadelphia: Fortress Press 1980, p. 73.

55. Moltmann, 'The Theology of Mystical Experience', p. 72.

56. R. S. Thomas, 'Journeys', *Mass for Hard Times*, Newcastle-upon-Tyne: Bloodaxe Books 1992, p. 28.

57. See Philip Sheldrake, *Spirituality and Theology: Christian Living and The Doctrine of God*, London: Darton, Longman & Todd/New York: Orbis Books 1998, especially pp. 24–31.

58. Michel de Certeau, *The Mystic Fable*, ET Chicago: University of Chicago Press 1992, p. 5, but see the complete Introduction, pp. 1–26.

59. de Certeau, *Mystic Fable*, p. 13.

60. de Certeau, *Mystic Fable*, p. 13.

61. de Certeau, *Mystic Fable*, p. 14.

62. A point noted by Luce Giard, one of de Certeau's closest collaborators and co-authors, in Michel de Certeau, Luce Giard and Pierre Mayol, *The Practice of Everyday Life*, volume 2, ET Minneapolis: University of Minnesota Press 1998, pp. xxii–xxiii.

63. de Certeau, *Mystic Fable*, p. 14.

64. de Certeau, *Mystic Fable*, p. 20.

65. See Michel de Certeau, 'The Weakness of Believing. From the Body to Writing, a Christian Transit', ET in Graham Ward (ed.), *The Certeau Reader*, Oxford: Blackwell 2000, *passim*.

66. Michel de Certeau, 'Culture and Spiritual Experience', *Concilium* 19 (1966), pp. 3–16.

67. I would put this more strongly than the 'conjectural' remarks of Frederick Bauerschmidt in his 'Introduction: Michel de Certeau, Theologian', in Ward, *Certeau Reader*, p. 213.

68. Karl Rahner, *The Spirit in the Church*. ET extracted in Geffrey Kelly (ed.), *Karl Rahner: Theologian of the Graced Search for Meaning*, Minneapolis: Fortress Press 1992, pp. 233–4.

69. In Paul Imhof and Hubert Biallowons (eds), *Karl Rahner in Dialogue*, New York: Crossroad 1986, pp. 196–7.

6. Re-Placing the City

1. These figures are cited by Sir Crispin Tickell in his Introduction to Richard Rogers, *Cities for a Small Planet*, London: Faber and Faber 1997, p. vii.

2. See, for example, the comments of Anne Buttimer in 'Home,

Reach and the Sense of Place', in Anne Buttimer and David Seamon (eds), *The Human Experience of Space and Place*, London: Croom Helm 1980.

3. Richard Sennett, *The Conscience of the Eye: The Design and Social Life of Cities*, London: Faber and Faber 1993, p. xii.

4. Sennett, *The Conscience of the Eye*, pp. xii–xiii.

5. Sennett, *The Conscience of the Eye*, pp. 6–10.

6. Sennett, *The Conscience of the Eye*, pp. 10–19.

7. See David Sibley, 'The Form of Purified Space', *Geographies of Exclusion*, London: Routledge 1997, pp. 78–81.

8. See Michel de Certeau, 'Practices of Space', in M. Blonsky (ed.), *On Signs*, Oxford: Blackwell 1985, pp. 122–45.

9. See Michael Northcott, 'A Place of Our Own?', in Peter Sedgwick (ed.), *God in the City: Essays and Reflections from the Archbishop of Canterbury's Urban Theology Group*, London: Mowbray 1995, pp. 119–38. Especially p. 122.

10. Arnold Berleant, *The Aesthetics of Environment*, Philadelphia: Temple University Press 1992, pp. 86–7.

11. Northcott, 'A Place of Our Own?', p. 122.

12. On the development of medieval cities see Jacques Le Goff, *Medieval Civilisation*, ET Oxford: Blackwell 1988, pp. 70–8.

13. See Peter Raedts, 'The Medieval City as a Holy Place', in Charles Caspers and Marc Schneiders (eds), *Omnes Circumadstantes: Contributions towards a History of the Role of the People in the Liturgy*, Kampen: Uitgeversmaatschappij J. H. Kok 1990, pp. 144–54.

14. For interesting remarks on the relationship between the fragmentation of intellectual discourse, starting with the medieval separation of theology and spirituality, and the contemporary secularization of the city, see James Matthew Ashley, *Interruptions: Mysticism, Politics and Theology in the Work of Johann Baptist Metz*, Notre Dame: University of Notre Dame Press 1998, pp. 10–12.

15. See Berleant, *The Aesthetics of Environment*, p. 62.

16. See Donlyn Lyndon and Charles W. Moore, *Chambers for a Memory Palace*, Cambridge, MA: MIT Press 1994, p. xii. The book has many stimulating thoughts about place throughout.

17. Marc Augé, *Non-Places: Introduction to an Anthropology of Supermodernity*, ET London/New York: Verso 1997, p. 66.

18. Augé, *Non-Places*, pp. 66–7.

19. Michael Novak, *The Spirit of Democratic Capitalism*, London: IEA Health and Welfare Unit 1991.

20. John Milbank, *Theology and Social Theory: Beyond Secular Reason*,

Oxford: Blackwell 1990, pp. 380–438.

21. Milbank, *Theology and Social Theory*, p. 403.

22. Milbank, *Theology and Social Theory*, p. 406.

23. For a critique of Milbank's approach to the human city, see Gillian Rose, 'Diremption of Spirit', p. 48, in Phillipa Berry and Andrew Wernick, *Shadow of Spirit: Postmodernism and Religion*, London: Routledge 1992.

24. Isidore of Seville, *Etymologiarum libri*, 15.2.1, quoted in Chiara Frugoni, *A Distant City: Images of Urban Experience in the Medieval World*, ET Princeton: Princeton University Press 1991, p. 3.

25. Augustine, *The City of God*, trans. Henry Bettenson, Harmondsworth: Penguin 1984.

26. See John S. Dunne, *The City of the Gods: A Study in Myth and Mortality*, London: Sheldon Press 1974, Chapter 7, 'The City of God', particularly p. 158.

27. This is the emphasis of Augustine's commentary on Genesis and is cited in Robert Markus, *The End of Ancient Christianity*, Cambridge: Cambridge University Press 1990, p. 78.

28. See the classic work on Augustine's theory of history, R. A. Markus, *Saeculum: History and Society in the Theology of St Augustine*, Cambridge: Cambridge University Press 1970, especially Chapter 1 'History: Sacred and Secular'.

29. Berleant, *The Aesthetics of Environment*, p. 172.

30. Michel de Certeau, 'Walking in the City', in his *The Practice of Everyday Life*. ET Berkeley: University of California Press 1988, pp. 91–110.

31. de Certeau, 'Indeterminate', in *The Practice of Everyday Life*, p. 203.

32. de Certeau, *The Practice of Everyday Life*, p. 115.

33. de Certeau, *The Practice of Everyday Life*, pp. 122–30.

34. See also the essay by Michael Northcott, 'Children', in Sedgwick (ed.), *God in the City*, pp. 139–52.

35. Franco LaCecia, 'The Space of Play', London: Royal Society of Arts, 1999.

36. It is disturbing that one of the largest, and apparently most comprehensive and interdisciplinary modern books on the city has no explicit reference at all to ethical or 'spiritual' perspectives and scarcely mentions religion except in terms of archaeology. This is explained by the surprising philosophical lacunae – no translations of continental thinkers, no references at all to figures such as de Certeau, Foucault or Augé and only

footnote citations of Lefebvre. See Richard T. LeGates and Frederic Stout (eds), *The City Reader*, London: Routledge 1999.

37. See Rogers, *Cities for a Small Planet*, p. 15.

38. Rogers, *Cities for a Small Planet*, p. 9.

39. Rogers, *Cities for a Small Planet*, pp. 167–8.

40. Haddon Wilmer, 'Images of the City and the Shaping of Humanity', in Anthony Harvey (ed.), *Theology in the City*, London: SPCK 1989, p. 37.

41. Peter Ackroyd, *The Life of Thomas More*, London: Random House 1999, p. 111.

42. From *Le Prêtre*, 1917, ET cited in the Introduction, p. 14, to Pierre Tielhard de Chardin, *Hymn of the Universe*, London: Collins 1965.

43. 'The Mass on the World', pp. 19–20, in Teilhard de Chardin, *Hymn of the Universe*.

Bibliography

Ackroyd, Peter, *The Life of Thomas More*, London: Random House 1999.

Appadurai, Arjun, *Modernity at Large: Cultural Dimensions of Globalisation*, Minneapolis: University of Minnesota Press 1998.

Aquinas, Thomas, *Summa Theologiae*.

Ashley, James Matthew, *Interruptions: Mysticism, Politics and Theology in the Work of Johann Baptist Metz*, Notre Dame: University of Notre Dame Press 1998.

Athanasius, *The Life of Anthony*, ET in Robert C. Gregg (ed.), *Athanasius: The Life of Anthony*, New York: Paulist Press 1980.

Augé, Marc, *Non-Places: Introduction to an Anthropology of Supermodernity*, ET London/New York: Verso 1997.

Augustine of Hippo, *The City of God*, trans. Henry Bettenson, Harmondsworth: Penguin 1984, and in *Nicene and Post-Nicene Fathers of the Christian Church*, volume II, Edinburgh: T&T Clark/Grand Rapids: Eerdmans 1993 edn.

—, 'Expositions on the Book of Psalms', Psalm XLII, in Philip Schaff (ed.), *A Select Library of the Nicene and Post-Nicene Fathers of The Christian Church*, Grand Rapids: Eerdmans/Edinburgh: T&T Clark 1996 edn, volume VIII.

—, *Homilies on the Psalms*, ET in Mary T. Clark (ed.), *Augustine of Hippo: Selected Writings*, London: SPCK/New York: Paulist Press 1984.

—, *Rule of St Augustine*, ET in George Lawless, *Augustine of Hippo and His Monastic Rule*, Oxford: Clarendon Press 1987.

Bachelard, Gaston, *The Poetics of Space*, Boston: Beacon Press 1994.

Baker, Denise Nowakowski, *Julian of Norwich's Showings*, Princeton: Princeton University Press 1994.

Bauerschmidt, Frederick Christian, *Julian of Norwich and the Mystical Body Politic of Christ*, Notre Dame: University of Notre Dame Press 1999.

Benedict of Nursia, *The Rule*, in Timothy Fry OSB (ed.), *The Rule of St. Benedict in Latin and English with Notes*, Collegeville, MN: Liturgical Press 1981.

Berleant, Arnold, *The Aesthetics of Environment*, Philadelphia: Temple University Press 1992.

Berry, Phillipa and Wernick, Andrew (eds), *Shadow of Spirit: Postmodernism and Religion*, London: Routledge 1992.

Bitel, Lisa, *Isle of the Saints: Monastic Settlement and Christian Community in Early Ireland*, Ithaca, New York: Cornell University Press 1990.

Blastic, Michael, 'Franciscan Spirituality', in Michael Downey (ed.), *The New Dictionary of Catholic Spirituality*, Collegeville, MN: Liturgical Press 1993.

Blumenfeld-Kosinski, Renate and Szell, Timea (eds), *Images of Sainthood in Medieval Europe*, Ithaca: Cornell University Press 1991.

Boff, Leonardo, 'The Need for Political Saints: From a Spirituality of Liberation to the Practice of Liberation', ET in *Cross Currents* XXX/4 (Winter 1980/81).

Bonaventure, *Bonaventure: The Soul's Journey into God; The Tree of Life; The Life of St Francis*, ed. Ewert Cousins, Classics of Western Spirituality, New York: Paulist Press 1978.

Bouyer, Louis, 'Mysticism – an Essay on the History of the Word', in Richard Woods (ed.), *Understanding Mysticism*, London: Athlone Press 1981, pp. 42–55.

Boyle, Marjorie O'Rourke, *Divine Domesticity: Augustine of Thagaste to Teresa of Avila*, Leiden: Brill 1997.

——, *Loyola's Acts: The Rhetoric of the Self*, Berkeley: University of California Press 1997.

Brink, André, *Looking on Darkness*, London: Secker & Warburg 1982.

Brown, David and Loades, Ann (eds), *The Sense of the Sacramental: Movement and Measure in Art and Music, Place and Time*, London: SPCK 1995.

Brown, Peter, *Authority and the Sacred: Aspects of the Christianisation of the Roman World*, Cambridge: Cambridge University Press 1995.

——, *The Body and Society: Men, Women and Sexual Renunciation in Early Christianity*, London: Faber and Faber 1991.

——, *The Cult of the Saints: Its Rise and Function in Latin Christianity*, London: SCM Press/Chicago: University of Chicago Press 1981.

——, *The Making of Late Antiquity*, Cambridge, MA: Harvard University Press, new edn. 1993.

——, 'The Notion of Virginity in the Early Church', in Bernard McGinn and John Meyendorff (eds), *Christian Spirituality: Origins to the Twelfth Century*, New York: Crossroad 1986.

—, *Society and the Holy in Late Antiquity*, Berkeley: University of California Press 1989.

Brueggemann, Walter, *Interpretation and Obedience: From Faithful Reading to Faithful Living*, Minneapolis: Fortress Press 1991.

—, *The Land: Place as Gift, Promise and Challenge in Biblical Faith*, Philadelphia: Fortress Press 1977; London: SPCK 1978.

Bultmann, Rudolph, *Jesus Christ and Mythology*, New York: Scribners 1958.

Burton-Christie, Douglas, 'Nature, Spirit and Imagination: The Quest for Sacred Place', unpublished lecture delivered at The College of San Rafael, California, September 1995.

—, *The Word in The Desert: Scripture and the Quest for Holiness in Early Christian Monasticism*, New York: Oxford University Press 1993.

Buttimer, Anne and Seamon, David (eds), *The Human Experience of Space and Place*, London: Croom Helm 1980.

Bynum, Caroline Walker, *Jesus as Mother: Studies in the Spirituality of the High Middle Ages*, Berkeley: University of California Press 1982.

Calvin, John, *Institutes of the Christian Religion*, trans. Henry Beveridge, repr. Grand Rapids: Eerdmans 1995.

Camille, Michael, *Gothic Art: Visions and Revelations of the Medieval World*, London: Weidenfeld & Nicolson 1996.

Candlin, Ann, 'Reflections on Elaine Graham's Model of Space', *Feminist Theology* 9 (May 1995), pp. 82–95.

Canter, David, *The Psychology of Place*, London: The Architectural Press 1977.

Carey, John, *The Faber Book of Utopias*, London: Faber and Faber 1999.

Carrette, Jeremy R., *Foucault and Religion: Spiritual Corporality and Political Spirituality*, London/New York: Routledge 2000.

Cavanaugh, William, 'The Eucharist as Resistance to Globalisation', in Sarah Beckwith (ed.), *Catholicism and Catholicity: Eucharistic Communities in Historical and Contemporary Perspectives*, Oxford: Blackwell 1999.

—, *Torture and the Eucharist: Theology, Politics and the Body of Christ*, Oxford: Blackwell 1998.

Certeau, Michel de, 'Culture and Spiritual Experience', *Concilium* 19 (1966).

—, Essays, in Graham Ward (ed.), *The Certeau Reader*, Oxford: Blackwell 2000.

—, 'How Is Christianity Thinkable Today?' ET in Graham Ward (ed.), *The Postmodern God*, Oxford: Blackwell 1997.

—, *The Mystic Fable*, ET Chicago: University of Chicago Press 1992.

—, *The Practice of Everyday Life*, ET Berkeley: University of California Press 1988.

—, 'Practices of Space', in M. Blonsky (ed.), *On Signs*, Oxford: Blackwell 1985, pp. 122–45.

Certeau, Michel de, Giard, Luce and Mayol, Pierre, *The Practice of Everyday Life*, volume 2, ET Minneapolis: University of Minnesota Press 1998.

Chadwick, Nora, *The Age of the Saints in the Early Celtic Church*, Oxford: Oxford University Press 1961.

Cloud of Unknowing, The, ed. James Walsh, New York: Paulist Press 1981.

Codina, Victor, 'Sacraments', pp. 218–19, in Jon Sobrino and Ignacio Ellacuria (eds), *Systematic Theology: Perspectives from Liberation Theology*, ET London: SCM Press 1996.

Cohen, Mark R., *Under Crescent and Cross: The Jews in the Middle Ages*, Princeton: Princeton University Press 1994.

Coleman, Simon and Elsner, John, *Pilgrimage: Past and Present in the World Religions*, London: British Museum Press/Cambridge, MA: Harvard University Press 1995.

Columbanus, *Sancti Columbani Opera*, ed. G. S. M. Walker, Scriptores Latini Hiberniae, vol. II, Dublin: The Dublin Institute for Advanced Studies 1970.

Cousins, Ewert, 'Franciscan Roots of Ignatian Mysticism', in George Schner (ed.), *Ignatian Spirituality in a Secular Age*, Waterloo, Ontario: Wilfred Laurier University Press 1984.

—, 'The Humanity and Passion of Christ', in Jill Raitt (ed.), *Christian Spirituality II: High Middle Ages and Reformation*, London: Routledge & Kegan Paul 1987, SCM Press 1989.

Cunningham, David S., *These Three Are One: The Practice of Trinitarian Theology*, Oxford: Blackwell 1998.

Curran, Charles, *The Church and Morality: An Ecumenical and Catholic Approach*, Minneapolis: Fortress Press 1993.

Dawes, E. and Baynes, N., *Three Byzantine Saints*, Oxford: B. Blackwell 1948.

Dionysius, *Pseudo-Dionysius: The Complete Works*, ed. Colm Luibheid, London: SPCK 1987.

Duby, Georges, *The Age of the Cathedral: Art and Society 980–1420*, ET

Chicago: University of Chicago Press 1981.

Dulles, Avery, *The Catholicity of the Church*, Oxford: Clarendon Press 1985.

—, *Models of the Church*, New York: Image Books 1987.

Dunne, John S., *The City of the Gods: A Study in Myth and Mortality*, London: Sheldon Press 1974.

Duns Scotus, John, *Opus Oxoniense*, II, 3, 6, 2.

Eade, John and Sallnow, Michael (eds), *Contesting the Sacred: The Anthropology of Christian Pilgrimage*, London: Routledge 1990.

Eco, Umberto, *Art and Beauty in the Middle Ages*, New Haven: Yale University Press 1986.

—, *Serendipities: Language and Lunacy*, London: Weidenfeld & Nicolson/ New York: Columbia University Press 1999.

Egeria, *Egeria's Travels*, trans. John Wilkinson, London: SPCK 1971.

Ekwall, Eilert, *The Concise Oxford Dictionary of English Place-Names*, Oxford: Clarendon Press 1985.

Eliade, Mircea, *Patterns in Comparative Religion*, New York: World Publishing 1963.

Ellacuria, Ignacio, 'The Church of the Poor, Historical Sacrament of Liberation', in Ignacio Ellacuria and Jon Sobrino (eds), *Mysterium Liberationis: Foundational Concepts of Liberation Theology*, ET New York: Orbis Books 1993, p. 543.

Elliott, Alison Goddard, *Roads to Paradise: Reading the Lives of the Early Saints*, Hanover: University Press of New England 1987.

Ellul, Jacques, *The Meaning of the City*, ET Carlisle: Paternoster Press 1997.

Faber, Hieje, *Above the Treeline: Towards a Contemporary Spirituality*, London: SCM Press 1988.

Feld, Steven and Basso, Keith H. (eds), *Senses of Place*, Santa Fe: School of American Research Press 1996.

Flanagan, Kieran, *The Enchantment of Sociology: A Study of Theology and Culture*, London: Macmillan/New York: St. Martin's Press 1999.

Ford, David, *Self and Salvation*, Cambridge: Cambridge University Press 1999.

Foucault, Michel, *Discipline and Punish: The Birth of the Prison*, ET Harmondsworth: Penguin, 1991.

—, *Power/Knowledge: Selected Interviews and Other Writings 1972–77*, ET London: Pantheon Books 1980.

Frances Teresa, *Living the Incarnation: Praying with Francis and Clare of Assisi*, London: Darton, Longman & Todd 1993.

Francis of Assisi, *Francis and Clare: The Complete Works*, eds. Regis Armstrong and Ignatius Brady, London: SPCK/New York: Paulist Press 1982.

Frugoni, Chiara, *A Distant City: Images of Urban Experience in the Medieval World*, ET Princeton: Princeton University Press 1991.

Fuller, Peter, *Theoria: Art and the Absence of Grace*, London: Chatto and Windus 1988.

Galilea, Segundo, 'Liberation as an Encounter with Politics and Contemplation', ET in Richard Woods (ed.), *Understanding Mysticism*, London: Athlone Press 1981.

—, 'The Spirituality of Liberation' *The Way* (July 1985), pp. 186–94.

Gallagher, Michael P., *Clashing Symbols: An Introduction to Faith and Culture*, London: Darton, Longman & Todd 1999.

Geertz, Clifford, *The Interpretation of Cultures*, New York: Basic Books, 1973.

Geremek, Bronislaw, 'The Marginal Man', in Jacques Le Goff (ed.), *The Medieval World*, ET London: Collins and Brown 1990.

Gibson, Katherine and Watson, Sophie (eds), *Postmodern Cites and Spaces*, Oxford: Blackwell 1995.

Gill, Robin, *Moral Communities: The Prideaux Lectures 1992*, Exeter: Exeter University Press 1992.

Graham, Elaine, 'From Space to Woman-Space', *Feminist Theology* 9 (May 1995), pp. 11–34.

Gray, Howard, 'What Kind of Document?', *The Way Supplement* 61 (Spring 1988).

Greinacher, Norbert, 'Catholic Identity in the Third Epoch of Church History', in James Prevost and Knut Walf (eds), *Concilium* 1994 no. 5, *Catholic Identity*, pp. 3–14.

Gunton, Colin, *The One, The Three and The Many*, Cambridge: Cambridge University Press 1995.

—, *The Promise of Trinitarian Theology*, Edinburgh: T&T Clark 1997.

Gurevich, A. J., *Categories of Medieval Culture*, ET London: Routledge & Kegan Paul 1985.

—, *Medieval Popular Culture: Problems of Belief and Perception*, ET Cambridge: Cambridge University Press 1990.

Gutiérrez, Gustavo, *We Drink from Our Own Wells*, ET London: SCM Press/New York: Orbis Books 1984.

Harbison, Peter, *Pilgrimage in Ireland*, London: Barrie & Jenkins 1991.

Harvey, Anthony (ed.), *Theology in the City*, London: SPCK 1989.

Hawley, John Stratton (ed.), *Saints and Virtues*, Berekeley: University of California Press 1987.

Heidegger, Martin, *Poetry,Language,Thought*, ET New York: Harper & Row 1975.

—, *The Question of Being*, ET New York: Twayne Publishers 1958.

Hopkins, Gerard Manley, *The Poems of Gerard Manley Hopkins*, eds W. H. Gardner and N. H. MacKenzie, Oxford: Oxford University Press 1989.

Hoskins, W.G., *The Making of the English Landscape*, London: Hodder & Stoughton revised edn 1992.

von Hügel, Freidrich, *The Mystical Element of Religion*, 1923, repr. London: James Clark 1927.

Hughes, Kathleen, *Church and Society in Ireland AD 400–1200*, London: Variorum 1987.

Hughes, Kathleen and Hamlin, Ann, *Celtic Monasticism: The Modern Traveller to the Irish Church*, New York: Seabury Press 1981.

Ignatius Loyola, *The Constitutions of the Society of Jesus*, ed. George Ganss, St Louis: The Institute of Jesuit Sources 1970.

—, *Ignatius of Loyola: Spiritual Exercises and Selected Works*, ed. George Ganss, New York: Paulist Press 1991.

—, *The Spiritual Exercises of Saint Ignatius*, ed.George Ganss, Chicago: Loyola University Press 1992.

Jantzen, Grace, *Julian of Norwich*, London: SPCK 1987.

—, *Power, Gender and Christian Mysticism,* Cambridge: Cambridge University Press 1995.

Joachim of Fiore, *The Book of Figures*, ET in Bernard McGinn (ed.), *Apocalyptic Spirituality – Treatises and Letters of Lactantius, Adso of Montier-en-Der, Joachim of Fiore, The Spiritual Franciscans, Savanarola*, New York: Paulist Press 1979, pp. 142–8.

Jones, L. Gregory and Buckley, James J. (eds), *Spirituality and Social Embodiment*, Oxford: Blackwell 1997.

Julian of Norwich, *A Book of Showings to the Anchoress Julian of Norwich*, eds Edmund Colledge and James Walsh, Montreal: Pontifical Institute of Mediaeval Studies 1978.

—, *Julian of Norwich: Showings*, eds Edmund Colledge and James Walsh, New York: Paulist Press 1978.

Kellcr, Catherine, *Apocalypse Now and Then*, Boston: Beacon Press 1996.

Kirshner, J. and Wemple, S. (eds), *Women of the Medieval World*, Oxford: Oxford University Press 1987.

LaCecia, Franco, 'The Space of Play', London: Royal Society of Arts 1999.

Lake, Carol, *Rosehill: Portraits from a Midlands City*, London: Bloomsbury 1989.

Lakoff, George and Johnson, Mark, *Metaphors We Live by*, Chicago: University of Chicago Press 1981.

Lane, Belden C., 'Galesville and Sinai: The Researcher as Participant in the Study of Spirituality and Sacred Space', *Christian Spirituality Bulletin* 2/1 (Spring 1994).

—, 'Simon Schama's *Landscape and Memory*', *Christian Spirituality Bulletin* 4/1 (Summer 1996).

—, *The Solace of Fierce Landscapes: Exploring Desert and Mountain Spirituality*, New York: Oxford University Press 1998.

Leenhardt, Franz, *Two Biblical Faiths: Protestant and Catholic*, ET London: SCM Press 1964.

Lefebvre, Henri, *The Production of Space*, ET Oxford: Blackwell 1991.

LeGates, Richard T. and Stout, Frederic (eds), *The City Reader*, London: Routledge 1999.

Le Goff, Jacques, 'Francis of Assisi between the Renewals and Restraints of Feudal Society', *Concilium* 149 (1981).

—, *Medieval Civilisation*, ET Oxford: Blackwell 1988.

—, *The Medieval Imagination*, ET London/Chicago: University of Chicago Press, 1988.

Lively, Penelope, *Spiderweb*, Harmondsworth/New York: Penguin 1999.

Louth, Andrew, *The Wilderness of God*, London: Darton, Longman & Todd 1991.

Lyndon, Donlyn and Moore, Charles W., *Chambers for a Memory Palace*, Cambridge, MA: MIT Press 1994.

McBrien, Richard, *Catholicism*, New York: HarperCollins 1994.

McDannell, Colleen and Lang, Bernhard, *Heaven: A History*, New Haven: Yale University Press 1988.

McElrath, D. (ed.), *Franciscan Christology*, New York: Franciscan Institute 1980.

McGinn, Bernard, *The Foundations of Mysticism: Origins to the Fifth Century*, 'The Presence of God. A History of Western Christian Mysticism', volume 1, New York: Crossroad 1992.

McManners, John (ed.), *The Oxford History of Christianity*, Oxford/New York: Oxford University Press 1993.

McNamara, Jo Ann Kay, *Sisters in Arms: Catholic Nuns through Two Millennia*, Cambridge, MA: Harvard University Press 1996.

Markus, Robert, *The End of Ancient Christianity*, Cambridge: Cambridge University Press 1990.

—, *Saeculum: History and Society in the Theology of St Augustine*, Cambridge: Cambridge University Press 1970.

Merton, Thomas, *Conjectures of a Guilty Bystander*, London: Sheldon Press 1977.

—, *Faith and Violence*, Notre Dame: University of Notre Dame Press 1968.

—, *Raids on the Unspeakable*, New York: New Directions 1966.

Miccoli, Giovanni, 'Monks', in Jacques Le Goff (ed.), *The Medieval World*, London: Collins & Brown 1990.

Milbank, John, *Theology and Social Theory: Beyond Secular Reason*, Oxford: Blackwell 1990.

Molinari, Paul, *Julian of Norwich: The Teaching of a Fourteenth Century English Mystic*, London: Burns & Oates 1958.

Moltmann, Jürgen, *Experiences of God*, Philadelphia: Fortress Press 1980.

Moore, R. I., *The Formation of a Persecuting Society*, Oxford: Blackwell 1994.

More, Thomas, *Utopia*, eds Edward Surtz and J. H. Hexter, The Complete Works of St Thomas More, volume 4, New Haven: Yale University Press 1965.

Morris, Richard, *Churches in the Landscape*, London: Dent 1989.

Mugerauer, Robert, *Interpretations on Behalf of Place: Environmental Displacements and Alternative Responses*, New York: State University of New York Press 1994.

Nadal, Jerome, *Monumenta Historica Societatis Jesu, Monumenta Nadal*, V, Institutum Historicum Societatis Jesu, Rome 1962.

Norris, Kathleen, *Dakota: A Spiritual Geography*, New York: Houghton Mifflin 1993.

Novak, Michael, *The Spirit of Democratic Capitalism*, London: IEA Health and Welfare Unit 1991.

Nuth, Joan, *Wisdom's Daughter: The Theology of Julian of Norwich*, New York: Crossroad 1991.

O'Leary, Brian, 'Living with Tension', *The Way Supplement* 61 (Spring 1988).

O'Malley, John, *The First Jesuits*, Boston, MA: Harvard University Press 1993.

Pannenberg, Wolfhart, *Christian Spirituality and Sacramental Community*, ET London: SCM Press 1984.

Panofsky, Erwin, *Abbot Suger on the Abbey Church of St Denis and Its Art Treasures*, Princeton: Princeton University Press 1979.

Pelikan, Jaroslav, *Christianity and Classical Culture*, New Haven: Yale University Press 1993.

Pieris, Aloysius, 'Spirituality and Liberation', *The Month* (April 1983).

Pohl, Christine D., *Making Room: Recovering Hospitality as a Christian Tradition*, Grand Rapids: Eerdmans 1999.

Raedts, Peter, 'The Medieval City as a Holy Place', in Charles Caspers and Marc Schneiders (eds), *Omnes Circumadstantes: Contributions towards a History of the Role of the People in the Liturgy*, Kampen: Uitgeversmaatschappij J. H. Kok 1990, pp. 144–54.

Raguin, Virginia Chieffo, Brush, Kathryn and Draper, Peter (eds), *Artistic Integration in Gothic Buildings*, Toronto: University of Toronto Press 1995.

Rahner, Hugo, *Ignatius the Theologian*, ET London: Geoffrey Chapman 1990.

Rahner, Karl, *Karl Rahner in Dialogue*, eds. Paul Imhof and Hubert Biallowons, New York: Crossroad 1986.

—, *The Spirit in the Church*, ET extracted in Geffrey Kelly (ed.), *Karl Rahner: Theologian of the Graced Search for Meaning*, Minneapolis: Fortress Press 1992.

—, 'The Theology of Mysticism', in K. Lehmann and L. Raffelt (eds), *The Practice of Faith: A Handbook of Contemporary Spirituality*, ET New York: Crossroad 1986, pp. 70–7.

—, *The Trinity*, ET London: Burns & Oates 1970.

Reeves, Marjorie, *Joachim of Fiore and the Prophetic Future*, London: SPCK 1976.

Ricoeur, Paul, *Figuring the Sacred: Religions, Narrative and Imagination*, ET Minneapolis: Fortress Press 1995.

—, *Time and Narrative*, 3 vols ET Chicago: University of Chicago Press 1984, 1985, 1988.

Rogers, Richard, *Cities for a Small Planet*, London: Faber and Faber 1997.

Rouet, Albert, *Liturgy and the Arts*, ET Collegeville, MN: Liturgical Press 1997.

Rule of the Master, The, ET Kalamazoo: Cistercian Publications 1977.

Ruusbroec, John, *John Ruusbroec: The Spiritual Espousals and Other Works*, ed. James Wiseman, New York: Paulist Press 1985.

Saliers, Donald E., 'Liturgy and Ethics: Some New Beginnings', in Ronald Hamel and Kenneth Himes (eds), *Introduction to Christian Ethics: A Reader*, New York: Paulist Press 1989, pp. 175–86.

Schama, Simon, *Landscape and Memory*, London: HarperCollins 1995.

Sedgwick, Peter (ed.), *God in the City: Essays and Reflections from the Archbishop of Canterbury's Urban Theology Group*, London: Mowbray, 1995.

Sennett, Richard, *The Conscience of the Eye: The Design and Social Life of Cities*, London: Faber and Faber 1993.

Sheldrake, Philip, *Living between Worlds: Place and Journey in Celtic Spirituality*, London: Darton, Longman & Todd/Boston: Cowley 1995.

—, *Spirituality and History: Questions of Interpretation and Method*, 2nd edn, London: SPCK 1995/New York: Orbis Books 1998.

—, *Spirituality and Theology: Christian Living and The Doctrine of God*, London: Darton, Longman & Todd/New York: Orbis Books 1998.

Sibley, David, *Geographies of Exclusion*, London: Routledge 1997.

von Simson, Otto, *The Gothic Cathedral: Origins of Gothic Architecture and the Medieval Concept of Order*, Princeton: Princeton University Press, expanded edn 1989.

Sobrino, Jon and Ellacuria, Ignacio (eds), *Systematic Theology: Perspectives from Liberation Theology*, ET London: SCM Press 1996.

Spohn, William, *Go and Do Likewise: Jesus and Ethics*, New York: Continuum 1999.

Teilhard de Chardin, Pierre, *Hymn of the Universe*, London: Collins 1965.

Thomas, Julian, *Time, Culture and Identity*, London: Routledge 1999.

Thomas, R. S., 'Journeys', *Mass for Hard Times*, Newcastle-upon-Tyne: Bloodaxe Books 1992.

Tilley, Christopher, *Metaphor and Material Culture*, Oxford: Blackwell 1999.

Tracy, David, *The Analogical Imagination: Christian Theology and the Culture of Pluralism*, SCM Press 1981, repr. New York: Crossroad 1991.

—, *On Naming the Present: God, Hermeneutics, and Church*, New York: Orbis Books 1994.

—, 'The Return of God in Contemporary Theology', *Concilium* 94 no. 6, *Why Theology?*

Traherne, Thomas, *Centuries*, London: Mowbray 1975.

Turner, Denys, *The Darkness of God: Negativity in Christian Mysticism*, Cambridge: Cambridge University Press 1995.

Turner, Victor and Edith, *Image and Pilgrimage in Christian Culture*, New York: Columbia University Press 1978.

Underhill, Evelyn, *Mysticism: The Nature and Development of Spiritual Consciousness*, 1911, repr. Oxford: One World Publications 1993.

Vauchez, André, *Sainthood in the Later Middle Ages*, ET Cambridge: Cambridge University Press 1997.

Veale, Joseph, 'How the Constitutions Work', *The Way Supplement* 61 (Spring 1988).

Verdeyen, Paul, *Ruusbroec and His Mysticism*, Collegeville, MN: Liturgical Press 1994.

Vivian, Tim (ed.), *Journeying into God: Seven Early Monastic Lives*, Minneapolis: Fortress Press 1996.

Ward, Benedicta (ed.), *The Wisdom of the Desert Fathers*, Oxford: Fairacres Publications 1986.

Ward, Graham, 'The Displaced Body of Jesus Christ', in John Milbank, Catherine Pickstock and Graham Ward (eds), *Radical Orthodoxy*, London: Routledge 1999, pp. 163–81.

Weil, Simone, *The Need for Roots*, ET London/New York; Routledge 1997.

Weinstein, Donald and Bell, Rudolph M., *Saints and Society: The Two Worlds of Western Christendom 1000–1700*, Chicago: University of Chicago Press 1982.

Whittock, Martyn, *Wiltshire Place-Names: Their Origins and Meanings*, Newbury: Countryside Books 1997.

Williams, Rowan, *On Christian Theology*, Oxford: Blackwell 2000.

—, 'Sacraments of the New Society', in David Brown and Ann Loades (eds), *Christ: The Sacramental Word*, London: SPCK 1996.

—, *Teresa of Avila*, London: Geoffrey Chapman 1991.

—, 'The Via Negativa and the Foundations of Theology: An Introduction to the Thought of V. N. Lossky', in Stephen Sykes and Derek Holmes (eds), *New Studies in Theology* 1, London: 1980.

Williams, Rowan and Sheldrake, Philip, 'Catholic Persons: Images of

Holiness. A Dialogue', in Jeffrey John (ed.), *Living the Mystery*, London: Darton, Longman & Todd 1994.

Wilson, Christopher, *The Gothic Cathedral*, London: Thames & Hudson 1990.

Wolter, Allan B. *The Philosophical Theology of John Duns Scotus*, Ithaca, New York: Cornell University Press 1990.

World Council of Churches, *Baptism, Eucharist and Ministry*, Faith and Order Paper 111, Geneva: World Council of Churches 1982.

Index